Lecture Notes in Artificial Intelligence 13739

Subseries of Lecture Notes in Computer Science

Series Editors

Randy Goebel
   University of Alberta, Edmonton, Canada

Wolfgang Wahlster
   DFKI, Berlin, Germany

Zhi-Hua Zhou
   Nanjing University, Nanjing, China

Founding Editor

Jörg Siekmann
   DFKI and Saarland University, Saarbrücken, Germany

More information about this subseries at https://link.springer.com/bookseries/1244

Neamat El Gayar · Edmondo Trentin ·
Mirco Ravanelli · Hazem Abbas (Eds.)

Artificial Neural Networks in Pattern Recognition

10th IAPR TC3 Workshop, ANNPR 2022
Dubai, United Arab Emirates, November 24–26, 2022
Proceedings

Editors
Neamat El Gayar
Heriot-Watt University
Dubai, United Arab Emirates

Edmondo Trentin
Universita' di Siena
Siena, Italy

Mirco Ravanelli
Concordia University
Montreal, QC, Canada

Hazem Abbas
Ain Shams University
Cairo, Egypt

ISSN 0302-9743   ISSN 1611-3349 (electronic)
Lecture Notes in Artificial Intelligence
ISBN 978-3-031-20649-8   ISBN 978-3-031-20650-4 (eBook)
https://doi.org/10.1007/978-3-031-20650-4

LNCS Sublibrary: SL7 – Artificial Intelligence

This Springer imprint is published by the registered company Springer Nature Switzerland AG
The registered company address is: Gewerbestrasse 11, 6330 Cham, Switzerland

Preface

This volume contains the papers presented at the 10th IAPR TC3 workshop on Artificial Neural Networks for Pattern Recognition (ANNPR 2022), held at Heriot-Watt University, Dubai, UAE, during November 24–26, 2022. ANNPR 2022 followed the success of the ANNPR workshops of 2003 (Florence), 2006 (Ulm), 2008 (Paris), 2010 (Cairo), 2012 (Trento), 2014 (Montreal), 2016 (Ulm), 2018 (Siena), and 2020 (Winterthur). The series of ANNPR workshops has acted as a major forum for international researchers and practitioners from the communities of pattern recognition and machine learning based on artificial neural networks.

Among the 24 submitted manuscripts, the Program Committee of the ANNPR 2022 workshop selected 16 papers for the scientific program, organized in regular oral presentations. Each paper was reviewed by at least two reviewers with a single-blind process. The workshop was enriched by two keynotes: "Advances and Paradigms in Deep Learning" given by Michael S. Lew, Leiden University, Netherlands, and "Evolutionary algorithms and neural networks: a match made in heaven?" given by Michael Lones, Heriot-Watt University, UK.

The workshop would not have been possible without the help of many people and organizations. First of all, we are grateful to all the authors who submitted their contributions to the workshop. We thank the members of the Program Committee and the many additional reviewers for performing the difficult task of selecting the best papers from a large number of high-quality submissions. The help of Kayvan Karim, from Heriot-Watt University in Dubai, who created and managed the ANNPR 2022 website, deserves a particular mention. Special thanks go to the School of Mathematical and Computer Sciences and the local organizing team from Heriot-Watt Dubai Campus for all the arrangements related to hosting ANNPR 2022 on their premises. ANNPR 2022 was supported by the International Association for Pattern Recognition (IAPR), the IAPR Technical Committee on Neural Networks and Computational Intelligence (TC3), and Heriot-Watt University, Dubai, UAE. Finally, we wish to express our gratitude to Springer for publishing our workshop proceedings within their LNCS/LNAI series. We hope that readers of this volume may enjoy it and get inspired by its contributions.

September 2022

Neamat El Gayar
Edmondo Trentin
Mirco Ravanelli
Hazem Abbas

Organization

ANNPR 2022 was organized by the School of Mathematical and Computer Sciences, Heriot-Watt University, Dubai, UAE.

Conference Chairs

Neamat El Gayar   Heriot-Watt University, UAE
Edmondo Trentin   Università di Siena, Italy
Mirco Ravanelli   Concordia Univesity, Canada
Hazem Abbas   Ain Shams University, Egypt

Program Committee

Shigeo Abe   Kobe University, Japan
Amir Atiya   Cairo University, Egypt
Erwin Bakker   Leiden University, Netherlands
Alessandro Betti   Université Côte d'Azur, France
Eugene Belilovsky   Concordia University, Canada
Jinde Cao   Southeast University, China
Ricardo Da Silva Torres Norwegian University of Science and Technology,
                           Norway
Renato De Mori   McGill University, Canada
Steven Ding   Queen's University, Canada
Andreas Fischer   University of Fribourg, Switzerland
Mohamed Gaber   Birmingham City University, UK
François Grondin   Université de Sherbrooke, Canada
Mohamed Hamdan   Yarmouk University, Jordan
Barbara Hammer   Bielefeld University, Germany
Mahmoud Khalil   Ain Shams University, Egypt
Michael Lones   Heriot-Watt University, UK
Wenlian Lu   Fudan University, China
Loren Lugosch   McGill University, Canada
Nadia Mana   Fondazione Bruno Kessler, Italy
Simone Marinai   University of Florence, Italy
Javier Montoya   HSLU Lucerne, Switzerland
Wei Pang   Heriot-Watt University, UK
Titouan Parcollet   Avignon Université, France
Marc Pouly   HSLU Lucerne, Switzerland

Simone Scardapane	Sapienza Università di Roma, Italy
Friedhelm Schwenker	Universität Ulm, Germany
Thilo Stadelmann	ZHAW, Switzerland
Cem Subakan	Université de Sherbrooke, Canada
Ah-Chung Tsoi	Macau University of Science and Technology, Macau
Md Azher Uddin	Heriot-Watt University, UAE

Local Arrangements

Kayvan Karim
Adrian Turcanu
Cristina Turcanu
Melita D'sa
Roselyne Okuoga

Sponsoring Institutions

International Association for Pattern Recognition (IAPR)
Technical Committee 3 (TC 3) of the IAPR
Heriot-Watt University, Dubai, UAE
Heriot-Watt University, Edinburgh, UK

Contents

Learning Algorithms and Architectures

Graph Augmentation for Neural Networks Using Matching-Graphs

Mathias Fuchs[1]([⊠]) [iD] and Kaspar Riesen[1,2] [iD]

[1] Institute of Computer Science, University of Bern, 3012 Bern, Switzerland
{mathias.fuchs,kaspar.riesen}@inf.unibe.ch
[2] Institute for Informations Systems, University of Applied Sciences Northwestern
Switzerland, 4600 Olten, Switzerland
kaspar.riesen@fhnw.ch

Abstract. Both data access and data collection have become increasingly easy over the past decade, leading to rapid developments in many areas of intelligent information processing. In some cases, however, the amount of data is still not sufficiently large (e.g. in some machine learning applications). Data augmentation is a widely used mechanism to increase the available data in such cases. Current augmentation methods are mostly developed for statistical data and only a small part of these methods is directly applicable to graphs. In a recent research project, a novel encoding of pairwise graph matchings is introduced. The basic idea of this encoding, termed matching-graph, is to formalize the stable cores of pairs of patterns by means of graphs. In the present paper, we propose to use these matching-graphs to augment training sets of graphs in order to stabilize the training process of state-of-the-art graph neural networks. In an experimental evaluation on five graph data sets, we show that this novel augmentation technique is able to significantly improve the classification accuracy of three different neural network models.

Keywords: Graph matching · Matching-graphs · Graph edit distance · Graph augmentation · GNN

1 Introduction and Related Work

Labeled training data is one of the most crucial requirements for the development and evaluation of supervised pattern recognition methods. This applies in particular to deep learning methods, which perform better the more examples of a given phenomenon a network is exposed to [1]. However, in real-world applications, we are often faced with the fact that the amount of training data is limited (for various reasons). Data augmentation techniques are a significant help with this problem. These techniques increase the amount of data by adding modified copies of the already existing entities to the training set. In the fields of computer vision [2] and natural language processing [3], for instance, data augmentation

Supported by Swiss National Science Foundation (SNSF) Project Nr. 200021_188496.

N. El Gayar et al. (Eds.): ANNPR 2022, LNAI 13739, pp. 3–15, 2023.
https://doi.org/10.1007/978-3-031-20650-4_1

is widely used. The vast majority of these augmentation algorithms are mainly suitable for statistical data (e.g. feature vectors) and not for structural data (e.g. graphs).

However, graphs can encode more information than merely an ordered list of real numbers. Hence, in terms of data representation, graphs are a versatile alternative to feature vectors. Due to their power and flexibility, graphs are widely used in the field of pattern recognition, and several graph distance models, graph kernels, and other techniques have been proposed in the literature [4]. Recently, *Graph Neural Networks* (GNNs) [5] have made great advances and are a sound option for data-driven graph algorithms [6,7]. Unfortunately, GNNs are very demanding when it comes to the amount of data they actually need. Therefore, GNNs might greatly benefit from data augmentation.

Due to the structural nature of graphs, there are comparatively few approaches to graph augmentation [8,9]. These rather few approaches mainly try to alter existing individual graphs by adding or removing edges. In [10], an approach is presented that is quite similar to our method. The authors propose an adaptation of the Mixup algorithm [11] for graphs. There are several neural network based approaches that are similar in spirit to our proposal, but not with the main goal of improving graph classification. Several approaches attempt to leverage the power of Variational Auto-Encoders (VAEs) for graph generation [12,13]. Furthermore, the success of Generative Adversarial Networks (GANs) for image generation led to an adaptation for graph generation as well [14,15].

In the present paper, we propose a novel approach for graph augmentation that differs to the previous methods in that we generate new graphs based on information captured in the matching of pairs of graphs. In particular, we exploit the so called *edit path* for augmentation purposes. The edit path gives us the information which subparts of the corresponding graphs are actually matched with each other by means of graph edit distance [16]. The graphs generated in this way are termed matching-graphs [17] and actually contain both edge and node modifications.

Our main hypothesis is that this novel method provides a natural way to create realistic and relevant graphs that actually provide a useful way to augment graph based data sets. Moreover, by extracting novel graphs from edit paths for any pair of training graphs available, the amount of training data can be increased virtually at any size.

The remainder of this paper is organized as follows. Section 2 provides an overview of the necessary theoretical background to make the paper self contained. In Sect. 3, we formally introduce the concept of matching-graphs, how they are built and used to augment a given training set of graphs. Eventually, in Sect. 4 we conduct an exhaustive experimental evaluation to provide empirical evidence that our approach of generating additional training samples is able to improve the classification accuracy of three different neural network architectures. Finally, in Sect. 5, we conclude the paper and discuss potential ideas for future work.

2   Theory and Basic Models

2.1   Graphs

Let L_V and L_E be finite or infinite label sets for nodes and edges, respectively. A *graph* g is a four-tuple $g = (V, E, \mu, \nu)$, where V is the finite set of nodes, $E \subseteq V \times V$ is the set of edges, $\mu : V \to L_V$ is the node labeling function, and $\nu : E \to L_E$ is the edge labeling function. This definition allows us to handle arbitrarily structured graphs with unconstrained labeling functions. In some algorithms it is necessary to include *empty "nodes"* and/or *empty "edges"*, both denoted by ε.

2.2   Graph Edit Distance (GED)

Given two graphs $g = (V, E, \mu, \nu)$ and $g' = (V', E', \mu', \nu')$, the basic idea of graph edit distance is to transform g into g' using some *edit operations*. A standard set of edit operations is given by *substitutions*, *deletions*, and *insertions*, of both nodes and edges. We denote the substitution of two nodes (and similarly edges) $u \in V$ and $v \in V'$ by $(u \to v)$, the deletion of node $u \in V$ by $(u \to \varepsilon)$, and the insertion of node $v \in V'$ by $(\varepsilon \to v)$.

A set $\{e_1, \ldots, e_s\}$ of s edit operations e_i that transform a source graph g completely into a target graph g' is called an *edit path* $\lambda(g, g')$ between g and g'. Typically, one assigns a cost of $c(e_i)$ to each edit operation e_i that reflects the strength of e_i. Intuitively, the graph edit distance is the minimum cost edit path between g and g'. Optimal algorithms for computing the edit distance of two graphs are typically based on combinatorial search procedures with exponential time complexity. Thus, applying graph edit distance to large graphs is computationally demanding (or not possible).

In order to reduce the computational complexity of graph edit distance, several approximation algorithms have been proposed in the literature [18,19]. In the present paper, we use the often employed approximation algorithm BP [20]. This specific algorithm reduces the problem of graph edit distance computation to an instance of a linear sum assignment problem for which several efficient algorithms exist. The approximated graph edit distance between g and g' computed by algorithm BP is termed $d_{\text{BP}}(g, g')$ from now on.

2.3   Graph Neural Networks (GNNs)

Of further relevance for the present paper is the concept of *Graph Neural Networks* [5]. GNNs provide a framework to use deep learning on graph structured data. The general goal of GNNs is to learn a vector h_v or h_g to represent each node $v \in V$ of a given graph or the complete graph g, respectively. That is, we obtain an embedding of either single nodes or complete graphs in a vector space. Based on this embedding, the individual nodes or the entire graph can then be classified. In our work, we focus on graph classification only.

In order to learn the graph embedding, we first need to learn the embedding of each node. In order to achieve this goal, GNNs usually follow a neighborhood aggregation strategy. That is, GNNs use a form of *neural message passing* in which messages are exchanged between the nodes of a graph [21]. The general idea is to iteratively update the representation of a node by aggregating the representation of its neighbors. During the i-th message-passing iteration, a hidden embedding $h_v^{(i)}$ that corresponds to each node $v \in V$ is updated according to the aggregated information from its neighborhood. Formally, this can be achieved by two differentiable functions.

- $m_{\mathcal{N}(v)}^{(i)} = AGGREGATE^{(i)}(\{h_u^{(i)} : u \in \mathcal{N}(v)\}$
- $h_v^{(i+1)} = UPDATE^{(i)}(h_v^{(i)}, m_{\mathcal{N}(v)}^{(i)})$

where $m_{\mathcal{N}(v)}^{(i)}$ is the "message" that is aggregated from the neighborhood $\mathcal{N}(v)$ of node v, and $\mathcal{N}(v)$ refers to the set of nodes adjacent to v. The feature vector representation of node v at the i-th iteration is $h_v^{(i)}$, and the initial representation $h_v^{(0)}$ is set to the vector $\mu(v)$ containing the original labels of v.

After N iterations of the message passing mechanism, this process produces node embeddings $h_v^{(N)}$ for each node $v \in V$. To get the embedding h_g for the complete graph, so-called *graph pooling* is necessary. Graph pooling combines the individual local node embeddings to one global embedding. That is, the pooling function maps the set of n node embeddings $\{h_{v_1}^{(N)}, \ldots, h_{v_n}^{(N)}\}$ to the graph embedding h_g. The pooling function can simply be a sum (or mean) of the node embeddings, or a more sophisticated function [22].

. In the present paper, we employ three different GNN architectures that are all based on the above mentioned concepts. The first architecture, denoted as *Simple* from now on, contains three Graph Convolutional Layers [7]. The second system is the Graph Isomorphism Network, denoted as *GIN*, introduced by Xu et al. [6]. The third architecture is the *GraphSAGE* network introduced by Hamilton et al. [23]. All three algorithms are implemented using Pytorch Geometric and for *GIN* and *GraphSAGE* we use the implementations of [24][1]. For the final graph classification, we add a Dropout Layer to all three architectures and feed the graph embedding into a fully connected layer.

3   Augment Training Sets by Means of Matching-Graphs

The main contribution of the present paper is to diversify and increase the size of a given training set of graphs. The rationale for this augmentation is that GNNs in particular rely on large training sets. We propose a novel method for graph augmentation based on so-called matching-graphs. Intuitively, a matching-graph is built by extracting information about the matching of pairs of graphs and formalising this information into a new graph. Matching-graphs can actually be interpreted as denoised core structures of the underlying graphs. In order

---

[1] https://github.com/diningphil/gnn-comparison.

to augment a given training set, the original definition of a matching-graph (introduced in [17]) is not suitable. Actually, matching-graphs in their original definition merely refer to subgraphs of the original graphs. Therefore, we propose an adapted definition for matching-graphs in the present context.

Formally, the process of creating matching-graphs can be described as follows. Given a pair of graphs $g = (V, E, \mu, \nu)$ and $g' = (V', E', \mu', \nu')$, the graph edit distance is computed first by means of algorithm BP [20]. This results in a (sub-optimal) edit path $\lambda(g, g') = \{e_1, \ldots, e_s\}$ that consists of s edit operations. Each node edit operation $e_i \in \lambda(g, g')$ can either be a substitution (denoted by $(u \rightarrow v)$, with $u \in V$ and $v \in V'$), a deletion (denoted by $(u \rightarrow \varepsilon)$, with $u \in V$), or an insertion (denoted by $(\varepsilon \rightarrow v)$, with $v \in V'$)[2].

Based on $\lambda(g, g')$ two matching-graphs $m_{g \times g'}$ and $m_{g' \times g}$ can now be built (one based on the source graph g and one based on the target graph g'). In order to create both $m_{g \times g'}$ and $m_{g' \times g}$, we initially define $m_{g \times g'} = g$ and $m_{g' \times g} = g'$. Then, we randomly select a certain percentage $p \in [0, 1]$ of all s edit operations available in $\lambda(g, g')$. Hence, we obtain a partial edit path $\tau(g, g') = \{e_1, \ldots, e_t\} \subseteq \lambda(g, g')$ with $t = \lfloor p \cdot s \rfloor$ edit operations only. Next, each edit operation $e_i \in \tau(g, g')$ is applied on graphs $m_{g \times g'}$ or $m_{g' \times g}$ according to the following rules.

- If e_i refers to a substitution $(u \rightarrow v)$, it is applied on both graphs $m_{g \times g'}$ and $m_{g' \times g}$. More precisely, the labels of the matching nodes $u \in V$ and $v \in V'$, are exchanged in both $m_{g \times g'}$ and $m_{g' \times g}$. Note that this operations shows no effect when the two labels of the involved nodes are identical.
- If e_i refers to a deletion $(u \rightarrow \varepsilon)$, e_i is applied on $m_{g \times g'}$ only, meaning that $u \in V$ is deleted in $m_{g \times g'}$.
- If e_i refers to an insertion $(\varepsilon \rightarrow v)$, e_i is applied on $m_{g' \times g}$ only. This means that the node $v \in V'$ that is inserted according to the edit operation, is deleted in $m_{g' \times g}$ instead.

The rationale for these rules is as follows. As stated above, the edge edit operations are always derived from the node edit operations. However, this derivation can only be performed if both nodes – adjacent to the edge – are actually edited. Hence, when inserting a node $v \in V'$, it is not necessarily clear how v should be connected to the remaining parts of the current graph. By applying the second and third rule above, this particular problem is avoided.

When all selected edit operations of $\tau(g, g')$ are applied on both graphs g and g', we finally obtain the two matching-graphs $m_{g \times g'}$ and $m_{g' \times g}$. Both matching-graphs represent intermediate graphs between the two underlying training graphs.

Clearly, if p is set to 1.0, $\tau(g, g')$ is equal to $\lambda(g, g')$, and thus all edit operations from the complete edit path are executed during the matching-graph creation. Note, however, that also in this case the corresponding matching-graphs differ from the original graphs due to the rules mentioned above. That is, deletions and insertions are uniquely applied to the source and the target graph,

---

[2] The edge edit operations are implicitly given by the node edit operations and thus not considered in $\lambda(g, g')$.

respectively. Hence, with $p = 1.0$, we obtain two matching-graphs that are subgraphs from the original graphs. For parameter values $0 < p < 1$, however, we obtain matching-graphs that are more diverse and particularly different from simple subgraphs (due to relabelled nodes).

Note that one can extract several partial edit paths $\tau(g, g')$ from one edit path $\lambda(g, g')$ using different values of p. This in turn results in several matching-graphs based on the same edit path. In theory, one can create as many matching-graphs as edit operations are available in $\lambda(g, g')$.

Figure 1 shows a visual example of the graph edit distance between two graphs g and g' and two possible resulting matching-graphs $m_{g \times g'}$ and $m_{g' \times g}$. The corresponding edit path is $\lambda = \{(0 \to a), (1 \to \varepsilon), (\varepsilon \to b), (2 \to c), (3 \to \varepsilon), (4 \to d)\}$. The possible matching-graphs $m_{g \times g'}$ and $m_{g' \times g}$, are created with $p = 0.5$, resulting in the partial edit path $\tau(g, g') = \{(0 \to a), (1 \to \varepsilon), (\varepsilon \to b)\}$, that consists of $t = 3$ edit operations. In this example it is clearly visible, that neither $m_{g \times g'}$ nor $m_{g' \times g}$ is a subgraph of g or g', respectively.

Note that the proposed process can lead to isolated nodes (as observed, for example, in the second last graph in Fig. 1). Based on the rationale that we aim to build graphs with nodes that are actually connected with each other, we remove unconnected nodes from the matching-graphs in our method.

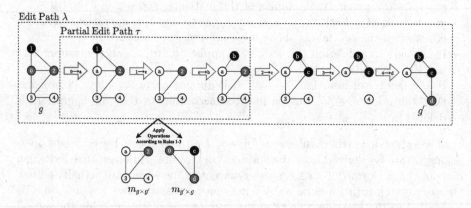

Fig. 1. An example of a complete edit path λ, a partial edith path τ, and the resulting matching-graphs $m_{g \times g'}$ and $m_{g' \times g}$.

Based on the process of creating two matching-graphs for pairs of graphs, we can now define an algorithm to augment a given training set with additional graphs. Algorithm 1 takes k sets of training graphs $G_{\omega_1}, \ldots, G_{\omega_k}$ stemming from k different classes $\omega_1, \ldots, \omega_k$ as input. The two **for** loops accomplish the following. For all pairs of graphs g, g' stemming from the same class ω_i, two matching-graphs $m_{g \times g'}$ and $m_{g' \times g}$ are built and added to the corresponding set of graphs (labeled with G_{ω_i}). Assuming n training graphs per class G_{ω_i} this results in $k \cdot n(n-1)$ matching-graphs in total, which are directly used to augment the corresponding training sets $G_{\omega_1}, \ldots, G_{\omega_k}$.

The percentage $p \in [0.1, 0.9]$ used for the creation of the matching-graph is randomly defined once for each iteration (see Line 5). Note, however, that inside the second **for** loop, just before the definition of p (Line 5), a further **for** loop could be defined, such that even more than one matching-graph could be created for each pair of graphs.

Algorithm 1: Graph Augmentation Algorithm

input : sets of graphs from k different classes $\mathcal{G} = \{G_{\omega_1}, \dots, G_{\omega_k}\}$
output: same sets augmented by matching-graphs

1 **foreach** *set $G_{\omega_i} \in \mathcal{G}$* **do**
2      $M = \{\}$
3   **foreach** *pair of graphs $g, g' \in G_{\omega_i} \times G_{\omega_i}$* **do**
4   Compute $\lambda(g, g') = \{e_1, \dots, e_s\}$
5   Randomly define p in $[0.1, 0.9]$
6   Define τ by selecting $\lfloor p \cdot s \rfloor$ edit operations from λ
7   Build both matching-graphs $m_{g \times g'}$ and $m_{g' \times g}$ according to τ
8          $M = M \cup \{m_{g \times g'}, m_{g' \times g}\}$
9   **end**
10      $G_{\omega_i} = G_{\omega_i} \cup M$
11 **end**

---

4 Experimental Evaluation

4.1 Experimental Setup

The overall aim of the following experimental evaluation is to answer the question, whether or not matching-graphs can be beneficially employed as an augmentation technique. In particular, we aim at verifying whether augmented sets help to increase the accuracy of neural network based classifiers. In order to answer this question, we evaluate the three GNN architectures, described in Sect. 2, with and without data augmentation. That is, the reference models are trained on the original training sets only (denoted by *Simple*, *GIN*, and *Graph-SAGE*), whereas our novel models are trained with matching-graphs added to the training sets (denoted as *Simple$_{mg}$*, *GIN$_{mg}$*, and *GraphSAGE$_{mg}$*).

In order to counteract uncontrolled randomness of neural network initializations, each experiment is repeated five times and the average accuracy is finally reported (we use the same seeds for both the reference approach and the augmented approach).

4.2 Data Sets

We evaluate the proposed approach on five data sets (three of them are molecule data sets and two are from other domains). All data sets are obtained from the TUDatasets[3] database [25].

---

[3] https://graphlearning.io.

- The first three data sets contain graphs that represent chemical compounds. By representing atoms as nodes and bonds as edges, graphs can actually represent chemical compounds in a lossless and straightforward manner. The *Mutagenicity* data set is split into two classes, one containing mutagenic compounds, and the other non-mutagenic compounds. The *NCI1* data set originates from anti-cancer screens and is split into molecules that have activity in inhibiting the growth of non-small cell lung cancer and those that have no activity. The third data set, *COX-2*, contains cyclooxygenase-2 (COX-2) inhibitors with or without in-vitro activities against human recombinant enzymes.
- *Cuneiform* script is one of the oldest writing systems in the world. This data set contains graphs that represent 29 different Hittie cuneiform signs, obtained from nine cuneiform tablets. Each cuneiform sign consists of tetrahedron shaped markings (wedges). The resulting graphs represent each wedge of the sign by four nodes, that are categorically labeled by their point as well as the type of the glyph. Additionally, we have continuous labels on the nodes that represent the spatial coordinates of the wedge [26].
- The fifth and last data set *Synthie* is an algorithmically created data set. It consists of four classes and the nodes contain 15 real-valued continuous attributes.

4.3 Validation of Metaparameters

For each iteration of the experiment, the corresponding data set is split into a random training, validation, and test set, with a 60:20:20 split. As we primarily aim at comparing three different network architectures, once with the default training set and once with an augmented training set, we do not separately tune the hyper parameters. Instead, we use the parametrization proposed in [24]. The optimizer used is Adam with a learning rate of 0.001, and we use the Cross Entropy as loss function for all three models.

All Models are trained for 200 epochs (except for *Mutagenicity* and *NCI1*, where we train for 50 epochs only, due to computational limitations arising from the large number of graphs in these data sets). The models that perform the best on the validation sets are finally applied to the test sets.

In order to construct the matching-graphs, we make use of algorithm BP, which approximates the graph edit distance. The cost for node and edge operations, as well as a weighting parameter $\gamma \in [0, 1]$ that is used to trade-off the relative importance of node and edge edit costs are often optimized [20]. However, for the sake of simplicity, we employ unit cost of 1.0 for deletions and insertions of both nodes and edges, and γ is taken from preliminary experiments [17].

4.4 Test Results and Discussion

In Table 1 we compare the mean classification accuracies of the three reference models with the augmented models using matching-graphs (obtained in five iterations).

Table 1. Classification accuracies of three models (*Simple*, *GIN* and *GraphSage*), compared to the same three models with augmented training sets (*Simple$_{mg}$*, *GIN$_{mg}$* and *GraphSage$_{mg}$*). Symbols $\textcircled{x}$ or $\bullet$ indicate a statistically significant improvement or deterioration in x of the five iterations when compared with the respective reference system (using a Z-test at significance level $\alpha = 0.05$).

Data Set	Simple	Simple$_{mg}$	GIN	GIN$_{mg}$	GraphSage	GraphSage$_{mg}$
Mutagenicity	79.1 ± 0.7	81.7 ± 1.1 ③	80.2 ± 0.9	81.2 ± 0.3 ①	77.6 ± 1.1	77.6 ± 0.7
NCI1	68.1 ± 1.8	71.8 ± 0.7 ⑤	74.0 ± 1.7	76.3 ± 0.8 ③	72.8 ± 0.4	73.1 ± 1.4 ①
COX-2	68.1 ± 4.0	70.9 ± 4.0 ①	73.4 ± 6.0	69.8 ± 1.2 ❷	68.8 ± 3.0	74.3 ± 3.0 ③
Cuneiform	62.7 ± 3.4	85.6 ± 1.3 ⑤	74.0 ± 4.1	82.9 ± 2.3 ④	49.3 ± 6.4	69.3 ± 5.3 ⑤
Synthie	73.9 ± 5.3	99.8 ± 0.5 ⑤	75.1 ± 6.1	95.9 ± 1.9 ⑤	48.0 ± 6.9	68.3 ± 2.3 ⑤

Overall we can observe that the augmentation process generally works well for all three models. On the simple GNN, our approach outperforms the corresponding reference system on all data sets and all iterations. In total, 19 out of the 25 improvements are statistically significant[4]. On the two data sets *Cuneiform* and *Synthie*, the improvements observed are quite impressive, with an increase of -23 and -26% points, respectively. Using the augmented training sets in conjunction with the *GIN* model, we obtain a higher mean accuracy compared to the reference system on four out of five data sets. In total the augmented model *GIN$_{mg}$* statistically significantly outperforms the reference system *GIN* in 13 out of 25 iterations. Finally, using *GraphSage* trained on the augmented sets, we again outperform the reference system on all five data sets according to the mean accuracy. While on the *Mutagenicity* data set no statistically significant improvement can be observed, for *Cuneiform* and *Synthie* our fitted models work particularly well, with 5 statistically significant improvements. Finally, it can also be seen that the standard deviation of the accuracies for each data set is almost always smaller for the augmented approach. Hence, we can conclude that our system becomes in general more robust and stable when compared with the reference system.

In Fig. 2 we show – as an example on the *Cuneiform* data set – the reference accuracies as well as the accuracies of our approach as bar charts for all five iterations (in light and dark gray, respectively). The iterations are ordered from the worst to the best performing reference accuracy. We can clearly see that our approach outperforms the reference systems for all iterations on all models.

---

[4] The statistical significance is computed via Z-test using a significance level of $\alpha = 0.05$.

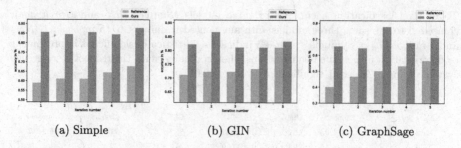

(a) Simple   (b) GIN   (c) GraphSage

Fig. 2. Classification accuracies of the five iterations on the *Cuneiform* data set using three different models (light bars) compared to our system that is based on the same models, but augments the training set with matching-graphs (dark bars).

Overall, the evaluation allows us to draw the following conclusion. Our approach is a viable graph augmentation technique and works well for all data sets and all models (except for the data set *COX-2* in conjunction with *GIN*). The augmentation leads to the least improvements on the *GIN* model, followed by *GraphSage* and finally *Simple* with the largest improvements. The differences of the accuracies on the *Simple* model is the most striking, suggesting that the augmentation approach with matching-graphs helps to bridge the gap when no sophisticated network architecture is available or applicable.

5   Conclusion and Future Work

The process of augmenting existing data to create larger amounts of training samples is especially important for systems that crucially depend on large data sets (like, for instance, neural networks). In the case of statistical data representations, quite an amount of methods are available for the task of data augmentation. Due to the complex nature of graphs, however, the research of graph augmentation is still behind its statistical counterparts.

In this work, we focus on augmenting and enlarging graph data sets for classification purposes. To this end, we propose to use so-called matching-graphs, which can be pre-computed by means of (sub-optimal) graph edit distance computations. Matching-graphs formalize the matching between two graphs by defining a novel graph that represents the core of each graph. This underlying core can be interpreted as a stable part of each graph. We systematically produce matching-graphs for each pair of training graphs and are thus able to substantially increase the amount of training data available. The goal of this augmentation process is to train more robust networks and finally improve the classification accuracy.

By means of an experimental evaluation of five graph data sets, we empirically confirm that our novel approach is able to improve three different graph neural network models in general. The vast majority of the improvements are statistically significant.

For future work we see different rewarding avenues. First, we feel that it could be beneficial to extend the definition of a matching-graph to include additional nodes and/or edges, to further increase the augmentation capability. Second, it would be interesting to see if the matching-graphs can be used in conjunction with other graph neural networks (e.g. Triplet loss networks or others).

References

1. Shorten, C., Khoshgoftaar, T.M.: A survey on image data augmentation for deep learning. J. Big Data **6**, 60 (2019). https://doi.org/10.1186/s40537-019-0197-0
2. Chinbat, V., Bae, S.H.: Ga3n: generative adversarial au-toaugment network. Pattern Recogn. **127**, 108637 (2022). https://doi.org/10.1016/j.patcog.2022.108637, https://www.sciencedirect.com/science/article/pii/S0031320322001182
3. Sahin, G.G., Steedman, M.: Data augmentation via dependency tree morphing for low-resource languages. In: Riloff, E., Chiang, D., Hockenmaier, J., Tsujii, J. (eds.) Proceedings of the 2018 Conference on Empirical Methods in Natural Language Processing, Brussels, Belgium, 31 October - 4 November 2018, pp. 5004–5009. Association for Computational Linguistics (2018). https://doi.org/10.18653/v1/d18-1545
4. Conte, D., Foggia, P., Sansone, C., Vento, M.: Thirty years of graph matching in pattern recognition. Int. J. Pattern Recognit Artif Intell. **18**(3), 265–298 (2004). https://doi.org/10.1142/S0218001404003228
5. Scarselli, F., Gori, M., Tsoi, A.C., Hagenbuchner, M., Monfardini, G.: The graph neural network model. IEEE Trans. Neural Netw. **20**(1), 61–80 (2009). https://doi.org/10.1109/TNN.2008.2005605
6. Xu, K., Hu, W., Leskovec, J., Jegelka, S.: How powerful are graph neural networks? In: 7th International Conference on Learning Representations, ICLR 2019, New Orleans, LA, USA, 6–9 May 2019. OpenReview.net (2019). https://openreview.net/forum?id=ryGs6iA5Km
7. Morris, C., et al.: Weisfeiler and leman go neural: higher-order graph neural networks. In: The 33rd AAAI Conference on Artificial Intelligence, AAAI 2019, The 31st Innovative Applications of Artificial Intelligence Conference, IAAI 2019, The 9th AAAI Symposium on Educational Advances in Artificial Intelligence, EAAI 2019, Honolulu, Hawaii, USA, 27 January - 1 February 2019, pp. 4602–4609. AAAI Press (2019). https://doi.org/10.1609/aaai.v33i01.33014602
8. Zhao, T., Liu, Y., Neves, L., Woodford, O.J., Jiang, M., Shah, N.: Data augmentation for graph neural networks. In: 35th AAAI Conference on Artificial Intelligence, AAAI 2021, 33rd Conference on Innovative Applications of Artificial Intelligence, IAAI 2021, The 11th Symposium on Educational Advances in Artificial Intelligence, EAAI 2021, Virtual Event, 2–9 February 2021, pp. 11015–11023. AAAI Press (2021). https://ojs.aaai.org/index.php/AAAI/article/view/17315
9. Zhou, J., Shen, J., Yu, S., Chen, G., Xuan, Q.: M-evolve: structural-mapping-based data augmentation for graph classification. IEEE Trans. Netw. Sci. Eng. **8**(1), 190–200 (2021). https://doi.org/10.1109/TNSE.2020.3032950
10. Park, J., Shim, H., Yang, E.: Graph transplant: node saliency-guided graph mixup with local structure preservation. In: 36th AAAI Conference on Artificial Intelligence, AAAI 2022, 34th Conference on Innovative Applications of Artificial Intelligence, IAAI 2022, The 12th Symposium on Educational Advances in Artificial Intelligence, EAAI 2022 Virtual Event, 22 February - 1 March 2022, pp. 7966–7974. AAAI Press (2022). https://ojs.aaai.org/index.php/AAAI/article/view/20767

11. Zhang, H., Cissé, M., Dauphin, Y.N., Lopez-Paz, D.: mixup: beyond empirical risk minimization. In: 6th International Conference on Learning Representations, ICLR 2018, Vancouver, BC, Canada, 30 April - 3 May 2018, Conference Track Proceedings. OpenReview.net (2018), https://openreview.net/forum?id=r1Ddp1-Rb

12. Kipf, T.N., Welling, M.: Variational graph auto-encoders. CoRR abs/1611.07308 (2016). arxiv.org/abs/1611.07308

13. Simonovsky, M., Komodakis, N.: GraphVAE: towards generation of small graphs using variational autoencoders. In: Kůrková, V., Manolopoulos, Y., Hammer, B., Iliadis, L., Maglogiannis, I. (eds.) ICANN 2018. LNCS, vol. 11139, pp. 412–422. Springer, Cham (2018). https://doi.org/10.1007/978-3-030-01418-6_41

14. Wang, H., et al.: Graphgan: graph representation learning with generative adversarial nets. In: McIlraith, S.A., Weinberger, K.Q. (eds.) Proceedings of the 32st AAAI Conference on Artificial Intelligence, (AAAI-18), the 30th innovative Applications of Artificial Intelligence (IAAI-18), and the 8th AAAI Symposium on Educational Advances in Artificial Intelligence (EAAI-18), New Orleans, Louisiana, USA, 2–7 February 2018, pp. 2508–2515. AAAI Press (2018). www.aaai.org/ocs/index.php/AAAI/AAAI18/paper/view/16611

15. Cao, N.D., Kipf, T.: Molgan: An implicit generative model for small molecular graphs. CoRR abs/1805.11973 (2018). arxiv.org/abs/1805.11973

16. Bunke, H., Allermann, G.: Inexact graph matching for structural pattern recognition. Pattern Recognit. Lett. 1(4), 245–253 (1983). https://doi.org/10.1016/0167-8655(83)90033-8

17. Fuchs, M., Riesen, K.: Matching of matching-graphs - a novel approach for graph classification. In: 25th International Conference on Pattern Recognition, ICPR 2020, Virtual Event/Milan, Italy, 10–15 January 2021, pp. 6570–6576. IEEE (2020). https://doi.org/10.1109/ICPR48806.2021.9411926

18. Chen, X., Huo, H., Huan, J., Vitter, J.S.: Fast computation of graph edit distance. CoRR abs/1709.10305 (2017). arxiv.org/abs/1709.10305

19. Fischer, A., Suen, C.Y., Frinken, V., Riesen, K., Bunke, H.: Approximation of graph edit distance based on hausdorff matching. Pattern Recognit. 48(2), 331–343 (2015). https://doi.org/10.1016/j.patcog.2014.07.015

20. Riesen, K., Bunke, H.: Approximate graph edit distance computation by means of bipartite graph matching. Image Vis. Comput. 27(7), 950–959 (2009). https://doi.org/10.1016/j.imavis.2008.04.004

21. Gilmer, J., Schoenholz, S.S., Riley, P.F., Vinyals, O., Dahl, G.E.: Neural message passing for quantum chemistry. In: International Conference on Machine Learning, pp. 1263–1272. PMLR (2017)

22. Ying, Z., You, J., Morris, C., Ren, X., Hamilton, W., Leskovec, J.: Hierarchical graph representation learning with differentiable pooling. In: Advances in Neural Information Processing Systems, vol. 31 (2018)

23. Hamilton, W.L., Ying, Z., Leskovec, J.: Inductive representation learning on large graphs. In: Guyon, I., (eds.) Advances in Neural Information Processing Systems 30: Annual Conference on Neural Information Processing Systems 2017, 4–9 December 2017, Long Beach, CA, USA, pp. 1024–1034 (2017). https://proceedings.neurips.cc/paper/2017/hash/5dd9db5e033da9c6fb5ba83c7a7ebea9-Abstract.html

24. Errica, F., Podda, M., Bacciu, D., Micheli, A.: A fair comparison of graph neural networks for graph classification. In: Proceedings of the 8th International Conference on Learning Representations (ICLR) (2020)

25. Morris, C., Kriege, N.M., Bause, F., Kersting, K., Mutzel, P., Neumann, M.: Tudataset: a collection of benchmark datasets for learning with graphs. In: ICML 2020 Workshop on Graph Representation Learning and Beyond (GRL+ 2020) (2020). www.graphlearning.io
26. Kriege, N.M., Fey, M., Fisseler, D., Mutzel, P., Weichert, F.: Recognizing cuneiform signs using graph based methods. In: International Workshop on Cost-Sensitive Learning, COST@SDM 2018, San Diego, California, USA, 5 May 2018. Proceedings of Machine Learning Research, vol. 88, pp. 31–44. PMLR (2018). http:// proceedings.mlr.press/v88/kriege18a.html

A Novel Representation of Graphical Patterns for Graph Convolution Networks

Marco Benini, Pietro Bongini, and Edmondo Trentin[✉]

DIISM - Università degli Studi di Siena, Siena, Italy
trentin@dii.unisi.it

Abstract. In the context of machine learning on graph data, graph deep learning has captured the attention of many researcher. Due to the promising results of deep learning models in the most diverse fields of application, great efforts have been made to replicate these successes when dealing with graph data. In this work, we propose a novel approach for processing graphs, with the intention of exploiting the already established capabilities of Convolutional Neural Networks (CNNs) in image processing. To this end we propose a new representation for graphs, called GrapHisto, in the form of unique tensors encapsulating the features of any given graph to then process the new data using the CNN paradigm.

Keywords: GrapHisto · Graph Neural Network · Graph Convolution Network

1 Introduction

In recent years Convolutional Neural Networks have achieved great results in processing and extracting informative features from grid-like data, such as images. Many researchers have tried to replicate this success on more complex and generic data. In particular, extracting high-level features representations on graph structured data have become a very attractive task. Graphs, indeed, can represent (in a concise yet explicative manner) a broad spectrum of relations among a variety of different entities. Likewise, they can capture relational properties underlying different portion of a dataset. However, generalizing existing deep learning techniques to this kind of data is very challenging: the model should be able to keep track of the structure of the graph as well as the attributes of nodes and edges and combine all the information in order to obtain meaningful and informative features of the data at hand. For these reasons many kinds of Graph Convolution Network (GCN) models have been proposed, such as Spectral-GCN or Spatial-GCNs, obtaining increasingly better results. Nevertheless, this kind of architectures often encounters similar difficulties, such as being constrained to certain kind of data or not being able to exploit all the characteristics in a graph, ending up in a loss of information. Two roads open up in this framework: trying to create better deep learning techniques for learning graph structured data or implementing clever solutions to provide graphs to already known architectures such as Convolutional Neural Networks (CNN). This paper aims to illustrate a novel method to represent graphs in a grid-like form in order to take advantage of the already established capacities of CNNs in extracting features. Following in the footsteps of image processing, this new kind of graph data representation, that we call GrapHisto, is our attempt in constructing an image able to describe all the features of a given graph at once in an effective and efficient way. In other words, the attractive and promising idea behind GrapHisto, is that they enable the passage from a graph processing problem into an equivalent, more intuitive, image processing problem.

GrapHistos were used to train a CNN in a classification task on a synthetic data-set and they proved to be a strong alternative to Graph Neural Network (GNN). In particular, a CNN with 2D convolutional layers, provided with the novel graph data representation as input, was tested on a classification task consisting in determining whether a graph is a tree or not. A GNN model was tested on the same task, obviously with its traditional data input. Multiple different choices of hyperparameters had been tried on each model with the intent of obtaining the best results on the task. The best achieved setups will be discussed as well as the comparison between each of them. The results show that the CNN achieve extremely better results then the GNN, suggesting that this novel representation of graphs in grid-like form is able to encapsulate the purely structural features of graphs in an effective way.

2   Related Work

Generally speaking, we refer to graph learning when we use machine learning on graphs. Different approaches rely on diverse, machine-readable representations of the input graphs. Due to the promising performances of deep learning in various fields, trying to extend the existing neural network models, such as recurrent neural networks (RNNs) or CNNs, to graph data has become an attractive task. Most of the state of the art methods are based on a representation of the graph in matrix form (adjacency or Laplacian) with the additional information of nodes and edge attributes still provided in a matrix form (often called "graph signal").

GNNs represent the state of the art [12,19]. The idea behind GNNs is simple: if we consider $\mathbf{x}$, $\mathbf{o}$, $\mathbf{l}$ and $\mathbf{l}_\mathcal{V}$ to be the vectors made by stacking respectively all the states $\mathbf{x}_n$, all the outputs, all the labels (of nodes and edges), and all the node labels, the computation can be formalized in terms of the following state equations: $\mathbf{x} = F_w(\mathbf{x}, \mathbf{l})$ and $\mathbf{o} = G_w(\mathbf{x}, \mathbf{l}_\mathcal{V})$, respectively, where the transition function F_w is responsible for the recursive update of node states and the output function G_w for the final encoding of graph structural and signal information. Although GNNs had are still widely used in various fields [12], they have some shortcomings, including: (i) GNNs are computationally expensive, since many iterations are needed to reach a steady state over cyclic graphs; (iii) they generally suffer from the vanishing gradient problem; (iii) F_w must be a contraction map in order to assure a unique solution to the state equations.

On the other hand, Spectral-GCNs rely on frequency filtering. Many Spectral-GCN models have been proposed since 2013 [5]. As in plain CNNs, [5] introduces a network with several spectral convolutional layers acting on the eigenvectors of the Laplacian matrix and the graph signal matrix. This method have a vertex domain localization problem, meaning that only the 1-hop neighbors features are considered. To address the issue, the ChebNet was proposed [7]. The ChebNet relies on new convolutional layers that use K-polynomial filters for localization, obtaining a good localization in the vertex domain. The computational complexity of the approach is limited by resorting to the Chebyshev expansion [7] to parameterize the filters with a polynomial function that can be computed recursively [10]. Other Spectral-GCNs were proposed to reduce the computational burden further, e.g. CayleyNet [18] (based on Cayley polynomials to define graph convolutions instead of using the Chebyshev expansion) and the Graph wavelet neural network (GWNN) [2] that replaces the Fourier transform in spectral filters by the graph wavelet transform.

Another drawback of Spectral-GCNs lies in their relying on the specific Laplacian matrix of any given graph, and in particular on its eigenfunctions. As a consequence, changing the input graph of such structures is not possible in general, since the eigenfunctions could be different. Spatial-GCNs [1,9,14] were proposed to address this issue by considering convolutions in the vertex domain, instead, working on some sort of aggregation among nodes signals within a certain neighborhood. The Neural Network for Graphs (NN4G) [15] represented the first attempt towards developing a Spatial-GCN. In NN4Gs, the graph convolutions work by summing up a node neighborhood information directly and repeat the process over multiple layers. More recent methods are based on the propagation and then aggregation of node representations from neighboring nodes in the vertex domain. Major instances are the Diffusion-GCN [1], the MoNet [9], and the SplineCNN [14]. These models do not account for possible edge attributes. This limitation was overcome by Simonovsky at al. [20], who implemented filters whose representative weights are conditioned on the specific edge features in the neighborhood of a node.

Grid-GCN [11,17] represent all the graph information in a grid-like form, taking into account the two main difference between grid-like data (e.g., bitmaps like in plain CNNs) and graph structured data. In order to use a grid-like representation of a generic graph, two issue shall be faced: (i) scanning a grid-like sample proceeds trivially from left to right and from top to bottom, while in graphs there is, in general, no such order; (ii) in a grid-like sample each node (e.g., pixel) has a fixed number of neighboring nodes, while general graphs has not. Suitable techniques for tackling (to some extent) these issues are found in the literature [11,17].

3   The GrapHisto

The proposed approach exploits a pseudo-visual representation (say, a "drawing") of graphs called GrapHisto. Even though it may be very helpful to draw graphs in order to infer their properties, the traditional node-link graph drawings present several shortcomings, including [3]:

1. There is no canonical representation: vertices and edges do not occupy regular, predictable locations. This is an instance of the classic graph matching problem.
2. While vertex labels may be put inside the circles representing the corresponding vertices, there is no clear and univocal place where to put the edge labels.
3. Lines representing edges intersect at points which are not vertices (any node-link drawing of a non-planar graph is affected by this shortcoming, as a consequence of Kuratowski theorem).
4. The drawing is visually dense: it may be hard to tell whether any given pair of vertices is connected by an edge, it may be difficult to tell whether certain pairs of vertices are actually connected, and it may be difficult to determine the degree of each vertex.
5. Since line segments representing edges can cross at very possible angles, it may be difficult to distinguish some pairs of edges from each other.

While [3] is concerned only with the purely graph-drawing viewpoint, we are herewith focusing on using the visual representation of graphs suitable to machine learning. As a consequence, the aforementioned limitations of the node-link representation are made more severe by the multidimensional nature of the labels (i.e., the feature vectors associated to nodes and/or edges). In order to introduce the GrapHisto, we need the following definition.

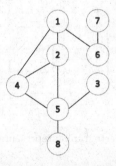

Fig. 1. A node-link representation of graph $\mathcal{H}$.

Definition 1. *Let* $\mathcal{G} = (\mathcal{V}, \mathcal{E})$ *be an undirected, loop-less graph and* $\mathcal{K}_{|\mathcal{V}|}$ *be the Kuratowski complete graph on the node set of* $\mathcal{G}$. *We define the* **Augmented Incidence Matrix** *of* $\mathcal{G}$ *to be the matrix* $I_{aug} \in \mathbb{R}^{|\mathcal{V}| \times \frac{|\mathcal{V}| \cdot (|\mathcal{V}|-1)}{2}}$ *such that* $I_{aug}(i, j) = 1$ *if and only if node* $i \in \mathcal{V}$ *is incident with edge* $j \in \mathcal{E}$. *That is:* I_{aug} *is the incidence matrix of* $\mathcal{K}_{|\mathcal{V}|}$ *with the columns corresponding to edges not present in* $\mathcal{E}$ *replaced by columns in which each entry is 0.*

Remark 1. Note that the augmented incidence matrix of a graph with loops is obtained in the exact same way but will have $\frac{|\mathcal{V}|\cdot(|\mathcal{V}|+1)}{2}$ columns, and the augmented incidence matrix of a digraph without loops will have $|\mathcal{V}| \cdot (|\mathcal{V}| + 1)$ columns.

Definition 2. *We define the* **Totally Augmented Incidence matrix** *of a an undirected loop-less graph* $\mathcal{G} = (\mathcal{V}, \mathcal{E})$ *to be the matrix* $I_{tot} \in \mathbb{R}^{|\mathcal{V}| \times \frac{|\mathcal{V}|\cdot(|\mathcal{V}|-1)}{2}}$ *such that* $I_{tot}(i, j) = 1$ *if and only if there exist* $k \leq i$ *and* $l \geq i$ *s.t* $I_{aug}(k, j) = 1$ *and* $I_{aug}(l, j) = 1$.

Remark 2. For directed graphs the Totally Augmented Incidence matrix can be defined similarly: simply replace 1 with -1 in the exact entries regarding destination nodes of particular edges. The rest can stay the same.

Now, starting from the Totally Augmented Incidence matrix of a graph we can easily construct the new graph drawing, referred to as the graph *cartographic representation* [3]. Suppose we want to obtain the cartographic representation of the graph $\mathcal{H}$ in Fig. 1:

Construction 1. *Take the Totally Augmented matrix of* $\mathcal{H}$ *and, in each column, substitute the consecutive entries that are equal to 1 with a vertical line. Replace the matrix with horizontal grey lines in place of the rows and vertical grey lines in place of the columns. The obtained drawing (see Fig. 2) is the* **cartographic representation** *of* $\mathcal{H}$.

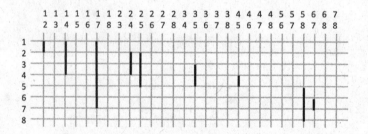

Fig. 2. The cartographic representation of $\mathcal{H}$.

Notice that the cartographic representation described above addresses issues 1–5 completely. However, since we are used to the classic node-link representation, it may be difficult to interpret this new drawings. In the cartographic drawing the nodes are represented by horizontal (grey) lines, while vertical black lines represent the edges. It is seen that (by construction) the latter cannot intersect with each other. Note that Construction 1 can be extended with some adjustments to directed graphs and to graph with loops. For graphs with loops the representation is exactly the same except the number of column is larger as mentioned in Remark 1. For directed graphs the author proposed the following construction:

Construction 2. *Take the Totally Augmented matrix of a directed graph $\mathcal{G}$; in each column substitute the consecutive entries that are equal to 1 with a vertical line and place a typographically attractive symbol in place of the $-1s$. Replace the matrix with horizontal grey lines in place of the rows and vertical grey lines on place of the columns. The obtained drawing is the cartographic representation of $\mathcal{G}$.*

As already mentioned, in the field of graph deep learning, we still have two more issues to address (namely, the need to account for multi-dimensional real-valued labels), but first we also have to adapt the cartographic representation to this new framework. From a computer vision point of view it is very easy to convert the new graph drawing to an image. We just have to consider the Totally Augmented matrix as a grid-like data sample in which each entry represents a particular pixel of an image. Figure 3 shows the cartographic representation of the sample graph $\mathcal{H}$ along with its associated image.

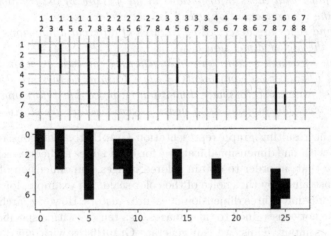

Fig. 3. The cartographic representation of $\mathcal{H}$ (top) and the associated image plotted by the matplotlib library (bottom).

Hereafter, it is beneficial to use the notion of tensor:

Definition 3. *(Tensor) A tensor T over K, of dimension n and type $I_1 \times I_2 \times \cdots \times I_n$ is a multidimensional table of elements of K, in which any element is determined by a multi index $(i_1, \ldots, i_n)$, where i_j ranges between 1 and I_j.*

When K is a field, such a tensor T is a **multilinear map**

$$T : K^{I_1} \times \cdots \times K^{I_n} \to K$$

where we consider the standard bases for each of the K^{I_i}. Now, it is seen that any given undirected acyclic graph $\mathcal{G} = (\mathcal{V}, \mathcal{E})$ can be stored as a three dimensional tensor T of type $|\mathcal{V}| \times \frac{|\mathcal{V}| \cdot (|\mathcal{V}|-1)}{2} \times I_3$, where I_3 depends on the dimension of the feature vectors of nodes and edges. We propose another construction to formalize the extension of graph cartographic representation for graph learning:

Construction 3. *Given an undirected loop-less graph $\mathcal{G} = (\mathcal{V}, \mathcal{E})$ construct its Totally Augmented Incidence matrix, without considering the features of edges and nodes. Then insert nodes and edges features as follows:*

- *if nodes and edges are described by features of the same dimension I_3:*
  1. *Consider a tensor T of dimension 3 and type $|\mathcal{V}| \times \frac{|\mathcal{V}| \cdot (|\mathcal{V}|-1)}{2} \times I_3$;*
  2. *label the mode-3 slices of T with indices from 0 to $I_3 - 1$;*
  3. *stack I_3 copies of the Totally Augmented Incidence matrix along mode-3 of tensor T;*
  4. *for each $i = 0, \dots, I_3 - 1$ and $k \in \mathcal{E}$ replace the ones representing edge k in the ith mode-3 slice with the ith component of edge k feature vector.*
  5. *for each $i = 0, \dots, I_3 - 1$ and $j = 0, \dots, |\mathcal{V}| - 1$ fill each entry (except the ones filled with edges information) of the jth row of the ith mode-3 slice with the ith component of node j feature vector.*
- *if nodes and edges are described by features of different dimensions:*
  1. *select the higher dimension I_3.*
  2. *if nodes features vectors have dimension I_3, proceed as before and skip step 4 whenever all components of edges features vectors are stored.*
  3. *if edges features vectors have dimension I_3, proceed as before and skip step 5 whenever all components of nodes features vectors are stored.*

We call the resulting graph representation GrapHisto. The construction of the graph $\mathcal{H}$ with one dimensional features for both nodes and edges is shown in Fig. 4. Notice that, in order to obtain colored images, the dimension of features vectors is constrained by the choice of the color scale. For example, for the RGB scale we can only have three dimensional features vectors. However, architectures like CNNs are not constrained to take images, i.e., tensors with a specific number of channels, as input. Thus, we can generate GrapHistos with any number of channels and still be able to feed them to a CNN. That being said, in Fig. 5 we exploit once again the potential of an image to show the GrapHisto representing graph $\mathcal{H}$ with three-dimensional features both for nodes and edges.

4   Preliminary Experimental Evaluation

Experiments were conducted on a synthetic dataset made of 20000 small graphs (9 vertices each). Half of them were trees and the other half presented at least one cycle. All graphs were generated using the library *NetworkX*[1] as follows. First, random trees having uniform distribution were generated via the `random_tree` function. Then, random cyclic graphs were generated starting from an Erdős-Rényi model (function `erdos_renyi_graph`) and selecting only the graphs presenting at least one cycle (function `find_cycle`).

---

[1] Available at https://networkx.org/.

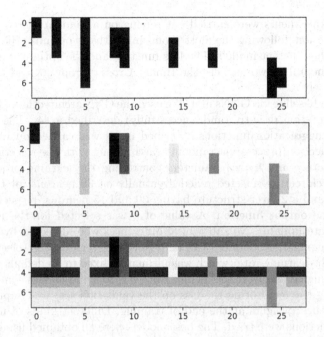

Fig. 4. From top to bottom: the image associated with the Totally Augmented Incidence matrix, the image after step *4* and the image after step *5*.

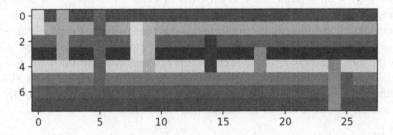

Fig. 5. GrapHisto of $\mathcal{H}$ with the addition of random 3D features for nodes and edges.

The learning task consisted in a plain binary classification task where the neural networks were trained to discriminate between trees and cyclic graphs. The performance of the resulting machines was assessed using the average Accuracy (ACC) and the average Area under the curve (AUC) metrics, ± the corresponding standard deviation. A 5-fold cross-validation strategy was applied, where for each fold as much as 60% of the data were used for training, 20% for validation, and 20% for test. Such a "shallow" investigation of the results is motivated by (i) the dramatic gap between the metrics yielded by the models, (ii) the negligible standard deviation obtained, and (iii) the large cardinality of the dataset which tends to prevent the bias due to random subsampling of the data.

All the experiments were carried out relying on a random search approach to model selection, following the most promising paths to optimize the hyperparameters of the different models. The loss function and the optimizer chosen for training the neural networks were the Binary Cross-entropy [6] and the Adam [13], respectively.

We fixed a baseline via GNNs first. We used the GNN architecture, algorithms and implementation recently (and, successfully) described in [4]. The choice of the different aggregation functions described in [4] was carried out as part of the model selection process via random search, along with the selection of the GNN architecture and hyperparameters controlling the learning process. The best GNN architecture selected relied eventually on 30 neurons for the transfer function, and a 2-layer structure having 30 and 15 neurons, respectively, for computing the output function. A value of 4 was selected for the maximum number of state updates. No substantial differences were detected by changing neither the aggregation modality (a plain "sum" aggregator was selected eventually), nor the learning rate (which was ultimately fixed to its default value, i.e. 0.001). Training epochs were set to 100. However, to avoid useless computation, early stopping (by monitoring the loss on the validation set) was implemented, restoring the best weights at the end of training. Different kinds of non linear activation functions were tried. The best models were all obtained using rectifier linear units (ReLUs). The GNN selected eventually involves an overall 3000 free parameters (approximately), that is substantially the same order of magnitude found in the selected GrapHisto-based GCNs (see below). Note that increasing the complexity of the GNN architecture (up to 30000 free parameters) did not yield any significant improvements in terms of cross-validated performance.

As for the GrapHisto-CNN, the model was implemented using the *Keras* library [8]. The different kinds of architectures tested were obtained by different combinations of the following types of layers (hereafter, each entry in the list is specified in terms of the keyword used by *Keras* followed in brackets by the acronym that we are using in the remaining of the paper):

- Conv2D (c): a 2-dimensional convolutional layer. In the present implementation, we have receptive field of size $(3,3)$ and stride set to 1
- MaxPooling2D (mp): 2-dimensional pooling layer
- GlobalMaxPooling (gmp) and GlobalAveragePooling (gap): extensions of mp to the global maximum and the global average, respectively
- Dense (ds): Fully connected feed forward layer

as well as of the following between-layers transformations of the signal were possibly applied:

- Flatten (f): flattens the input into a one dimensional vector.
- Dropout (dr): applies dropout as described in [16].

Hereafter, we are using strings in the form "c-mp-dr-..." in order to represent a CNN architecture consisting in a 2-dim convolutional layer followed by a 2-dim pooling layer, followed by an application of dropout, etc. Table 1 reports the characteristics (architectures and hyperparameters) of the different GrapHisto-GCN variants selected. We refer to these variants as $GH_1, \ldots, GH_4$, respectively. All the convolutional layers in these GCNs rely on receptive fields set to (3,3) and stride set to (1,1). The column labeled "Conv. filters" in the Table specifies the number of filters within each Conv2D layer, in the order and separated by commas. For instance, if the architecture includes 3 convolutional layers (say, "...-c-...-c-...-c-..."), the corresponding "Conv. filters" column will contain a triplet n_1, n_2, n_3 where n_1 is the number of filters in the first Conv2D layer, n_2 the number of filters in the second, and so on. Next, the column labeled "Pool. dim." reports the dimension of the receptive fields for the max-pooling (they all turned out to be equal in the present setup, upon application of the model selection procedure). The column "Drop" specifies the percentages of dropping-out that were selected for the dropout (in the order, and separated by commas, for each and every occurrence of "dr" within the architecture descriptor "...-dr-...-dr-..."). Finally, the column "Dense" reports the number of neurons in each fully-connected feed-forward layer found in the architecture ("...-ds-...-ds-..."), in the order and comma-separated. The selected GCNs embraced the following approximate number of free parameters: 3000 (GH_1), 6000 (GH_2), 14000 (GH_3), and 1800 (GH_4), respectively. More complex architectures suffered from the phenomenon of overfitting the training data, and they ended up worsening their generalization capabilities to some extent.

Table 1. Characteristics of the different GrapHisto-GCN variants selected.

Model	Architecture	Conv. filters	Pool. dim	Drop	Dense
GH_1	c-c-mp-dr-c-mp-dr-gmp-ds-ds	8,16,16	(2,2)(2,2)	0.05,0.13	16,1
GH_2	c-c-mp-c-mp-dr-gmp-ds-ds	8,16,32	(2,2)(2,2)	0.08, 0.20	16,1
GH_3	c-c-mp-dr-c-mp-dr-gmp-ds-ds	16,32,32	(2,2)(2,2)	0.13, 0.3	32,1
GH_4	c-c-mp-c-mp-dr-gmp-ds-ds	4,8,16	(2,2)(2,2)	0.10	16,1

The results obtained are presented in Table 2 in terms of average AUC and ACC ($\pm$ standard deviation) over the 5-fold evaluation procedure. The GCNs previously selected are compared with respect to each other, as well as w.r.t. the baseline yielded by the GNN (the state-of-the-art paradigm for the classification of graphical patterns). It is seen that: (i) the proposed approach outperforms the GNN in the present classification task. The gap (in terms of both AUC and ACC) is so dramatic that no explicit statistical significance test is actually needed (also in the light of the extremely limited standard deviations observed); (ii) all the different GCN variants yield significantly good performance in the task, corroborating the effectiveness of the proposed approach; (iii) differences among the different GCNs in terms of AUC and ACC are very limited, especially in consideration of the gap over the GNN and of the limited standard deviations, proving the approach to be stable and little sensitive to choices pertaining the architecture and the hyperparameters; (iv) stability of the approach is all the more evident if the standard deviation are considered.

Table 2. Results yielded by the different GCN variants (5-fold, *avg.* $\pm$ *std. dev.*), compared with the best GNN. The highest values of AUC and ACC obtained are printed in boldface.

Model	AUC	ACC
GNN	$93.32_{\pm 0.24}\%$	$86.02_{\pm 0.33}\%$
GH_1	$98.01_{\pm 0.13}\%$	$95.98_{\pm 0.19}\%$
GH_2	$\mathbf{98.12}_{\pm 0.24}\%$	$\mathbf{96.20}_{\pm 0.33}\%$
GH_3	$97.34_{\pm 0.18}\%$	$95.11_{\pm 0.34}\%$
GH_4	$98.07_{\pm 0.46}\%$	$96.06_{\pm 0.26}\%$

5   Conclusion

Classification of graphical patterns is a relevant task to many applications (in bioinformatics, cheminformatics, semantic networks, etc.). Several neural networks for graphs have been proposed. Most popular instances are the GNN and several variants of the GCN. The paper introduced GrapHisto, a novel representation of graphs that is suitable for processing via CNNs. GrapHisto is a CNN-oriented, multi-channel extension to the cartographic representation of graphs. Preliminary experimental results on a large synthetic 2-class graph classification task corroborated the viability of the approach, its stability, and its potentialities w.r.t. the state-of-the-art machines.

References

1. Atwood, J., Towsley, D.: Diffusion-convolutional neural networks. In: Advances in Neural Information Processing Systems, pp. 1993–2001 (2016)
2. Xu, B., et al.: Graph wavelet neural network. arXiv preprint. arXiv:1904.07785 (2019)
3. Blakley, B., Blakley, G., Blakley, S.: How to draw graphs: Seeing and redrafting large networks in security and biology. CoRR abs/1405.5523 (2014)
4. Bongini, P., Bianchini, M., Scarselli, F.: Molecular generative graph neural networks for drug discovery. Neurocomputing **450**, 242–252 (2021)
5. Bruna, J., Zaremba, W., Szlam, A., LeCun, Y.: Spectral networks and locally connected networks on graphs. arXiv preprint. arXiv:1312.6203 (2013)
6. Cox, D.: The regression analysis of binary sequences. J. Roy. Stat. Soc.: Ser. B (Methodol.) **20**(2), 215–232 (1958)
7. Defferrard, M., et al.: Convolutional neural networks on graphs with fast localized spectral filtering. In: 29th NIPS, pp. 3844–3852 (2016)
8. Chollet, F., et al.: Keras (2015). https://github.com/fchollet/keras

9. Monti, F., et al.: Geometric deep learning on graphs and manifolds using mixture model cnns. CoRR abs/1611.08402 (2016)
10. Hammond, D.K., et al.: Wavelets on graphs via spectral graph theory. Appl. Comput. Harmon. Anal. **30**(2), 129–150 (2011)
11. Gao, H., et al.: Large-scale learnable graph convolutional networks. In: Proceedings of the 24th ACM SIGKDD, pp. 1416–1424 (2018)
12. Zhou, J., et al.: Graph neural networks: a review of methods and applications. AI Open **1**, 57–81 (2020)
13. Kingma, D., Ba, J.: Adam: a method for stochastic optimization (2017)
14. Fey, M., et al.: Splinecnn: fast geometric deep learning with continuous b-spline kernels. CoRR abs/1711.08920 (2017)
15. Micheli, A.: Neural network for graphs: a contextual constructive approach. IEEE Trans. Neural Netw. **20**(3), 498–511 (2009)
16. Srivastava, N., et al.: Dropout: a simple way to prevent neural networks from overfitting. JMLR **15**(56), 1929–1958 (2014)
17. Niepert, M., et al.: Learning convolutional neural networks for graphs. In: Proceedings of The 33rd ICML, vol. 48, pp. 2014–2023 (2016)
18. Levie, R., et al.: Cayleynets: graph convolutional neural networks with complex rational spectral filters. CoRR abs/1705.07664 (2017)
19. Scarselli, F., et al.: The graph neural network model. IEEE Trans. Neural Netw. **20**(1), 61–80 (2009)
20. Simonovsky, M., Komodakis, N.: Dynamic edge-conditioned filters in convolutional neural networks on graphs. In: Proceedings of the IEEE Conference on Computer Vision and Pattern Recognition, pp. 3693–3702 (2017)

Minimizing Cross Intersections in Graph Drawing via Linear Splines

Rida Ghafoor Hussain[1(✉)], Matteo Tiezzi[2], Gabriele Ciravegna[3], and Marco Gori[2,3]

[1] University of Florence, Florence, Italy
rida.ghafoor@unifi.it
[2] University of Siena, Siena, Italy
{mtiezzi,marco.gori}@unisi.it
[3] Université Cóte d'Azur, Nice, France
gabriele.ciravegna@inria.fr

Abstract. The generation of aesthetically pleasing graph layouts is the main purpose of Graph Drawing techniques. Recent contributions delved into the usage of Gradient-descent (GD) based schemes to optimize differentiable loss functions, built to measure the graph layout adherence to given layout characteristics. However, some properties cannot be easily expressed via differentiable functions. In this direction, the recently proposed Graph Neural Drawer (GND) framework proposes to exploit the representational capability of neural models in order to be able to express differentiable losses, specifically for edge intersection, that can be subsequently optimized via GD. In this paper, we propose to improve graph layout readability leveraging linear splines. We exploit the principles behind GND and use a neural model both to identify crossing edges and to optimize their relative position. We split crossing edges introducing linear splines, and threat the control points as novel "fake" vertices that can be optimized via the underlying layout optimization process. We provide qualitative and quantitative analysis over multiple graphs and optimizing different aesthetic losses, that show how the proposed method is a viable solution.

Keywords: Graph Drawing · Gradient Descent · Graph Neural Drawers · Splines

1 Introduction

The widening interest over structured data and its representation calls to the development of methods and techniques to better visualize and investigate its complex relations and characteristics [8]. Graphs are mathematical structures which model pairwise relations (edges) between entities (nodes). Graph Drawing [5] methods focus on developing algorithmic techniques to generate drawings of graphs via, for instance, the node-link paradigm [10], i.e. entities represented as nodes and their relation expressed by edges linking them.

N. El Gayar et al. (Eds.): ANNPR 2022, LNAI 13739, pp. 28–39, 2023.
https://doi.org/10.1007/978-3-031-20650-4_3

Fig. 1. Reducing the number of arc intersections via linear splines. On the left, a Diamond Graph layout has been obtained via an optimization process that is not able to fully avoid arc intersections. The proposed method introduce linear splines, whose control points (*fake nodes*), plotted with an alpha color transparency, are treated as additional nodes for the underlying optimization process. This approach ease the graph drawing optimization process.

The prominent role of graph drawing is the improvement of graph readability, which is generally evaluated with some aesthetic criteria such as community preservation, crossing angles, edge length variance, number of crossing edges, etc. [2]. The final goal is to devise appropriate coordinates for the node positions, a goal that often requires to explicitly express and combine the aforementioned criteria. Beyond the classic Graph Drawing methods that are based on energy models [15,16] or spring-embedders [7,12], interesting directions are the ones which try to express the graph aesthetics via a differentiable function that can be optimized via Gradient Descent [2,29]. Machine learning applications have been used also to inject the user preferences into the layout optimization process. Such methods exploit evolutionary algorithms (e.g. genetic algorithms) to learn user preferences [3,4] keeping the human interaction in loop – causing however an inherent dependence on the user. Lately, given the increasing successes of Deep Learning models in several research fields, several works applied these architectures into the field of Graph Drawing. Neural networks are capable to learn the layout characteristics from graph drawing techniques [25] or from the layout distribution itself [17]. A recent contribution from Tiezzi et al. [21] introduced the general framework of Graph Neural Drawers (GNDs), where the main intuition is that neural networks (i.e., *Neural Aesthetes*) can be used both to express in a differentiable-manner the cross-intersection among arcs and to use this information as a loss that can be optimized via Graph Neural Networks (GNNs) [20,22,23]. In particular, the pre-trained *Neural Aesthete* exploited in the first step predicts if any two edges cross. Then, since neural networks outputs are differentiable, the well-trained edge crossing predictor serves as a guide to gradient descent steps to move the node towards the direction of non-intersection.

We build on this principles, introducing a novel method to obtain better and more pleasing layouts where the number of intersections is reduced. We propose to use a *Neural Aesthete* to devise if two arcs are crossing, and, if that is the case, we split one of the arc introducing a linear spline. The obtained control points of the spline can be threated as additional nodes of the graph, what we refer to as a *fake vertex*, that is added to the layout optimization process. This idea gives further freedom to the layout generation dynamics, with the final goal to move the newly obtained segments (i.e., the fake node coordinates) into direction where the number intersection is highly reduced. We depict this intuition in Fig. 1. We prove that this approach can be further extended not only in optimizing the number of intersecting arcs, but also with other relevant layout losses, such as the *stress function* [29]. Our experimental findings show that our proof-of-concept is a viable solution for several graphs. An identifiable decrease in number of cross intersection of arcs is observed, overal several input graphs we tested, after the concept of linear splines is introduced into the optimization process.

The paper is organized as follows. Section 2 introduces some references on the Graph Drawing literature. Section 3 describe the concepts of Neural Aesthete and how to use it to introduce linear splines into the graph layout. Section 4 describes our proof-of-concept and the obtained experimental findings. Conclusions are drawn in Sect. 5.

2   Related Work

Literature provides a large variety of research methods aimed at improving graph readability. Most of these algorithms are designed to optimize a single aesthetic criteria. The main directions, depicted in the relevant survey by Gibson et al. [13] (i.e. force-directed, dimension reduction, multi-level techniques), involve the generation of interesting and aesthetically pleasing layouts by methods which regard a graph as a physical system, with forces acting on nodes with attraction and repulsion dynamics up to a stable equilibrium state [16]. For instance, a classic layout criterion is brought by the minimization of the *stress* function [16], where the node positions are computed in such a way that the actual distance between node pairs gets proportional to their graph theoretical distance. A very effective approach that have been proved to enhance the human understanding of the layouts and graph topologies, consists in reducing the number of cross intersections in edges [19]. However, the computational complexity of the problem is NP-hard, and several authors proposed complex solutions and algorithms to address this problem [1].

Recently, several works moved towards the direction of improving multiple layout criteria at once. Wang et al. [26] propose a revised formulation of stress that can be used to specify ideal edge direction in addition to ideal edge lengths. Devkota et al. [9] also use a stress-based approach to minimize edge crossings and maximize crossing angles. Eades et al. [11] provided a technique to draw large graphs while optimizing different geometric criteria, including the Gabriel graph property. Such approaches are capable to optimize multiple aesthetic criteria, but they cannot adapt naturally and dynamically to handle further optimization goals. Some recent contributions in this specific context presented the advantages of applying gradient based methodologies in graph drawing tasks. Zheng et al. [28] efficiently applied the Sthocastic Gradient Descent (SGD) method to reduce the stress loss, displacing pairs of vertices following the direction of the gradient. The recent framework by Ahmed et al. [2], $(GD)^2$, leverages

Gradient Descent to optimize several readability criteria at once, as long as the criterion can be expressed by smooth functions. Indeed, thanks to the powerful auto-differentiation tools available in modern machine learning frameworks [18], several criteria such as ideal edge lengths, stress, node occlusion, angular resolution and many others can be optimized smoothly and with ease. Finally, we report some early attempts to leverage Deep Learning Models in the Graph Drawing scenario. Wang et al. [26] proposed a graph-based LSTM model able to learn and generalize the coordinates patterns produced by other graph drawing techniques on certain graph layouts. Very recently, in DeepGD [24] a message-passing Graph Neural Network (GNN) is leveraged to process starting positions produced by graph drawing frameworks [6] to construct pleasing layouts that minimize combinations of aesthetic losses (stress loss combined with others) on arbitrary graphs. The Graph Neural Drawer framework [21] focus on the criteria of edge crossing, proving that a Neural Network (regarded as a *Neural Aesthete*) can learn to identify if two arcs are crossing or not. This simple model provides a useful and flexible gradient direction that can be exploited by (Stochastic) Gradient Descent methods. Moreover, the GND framework proved that GNNs, even in the non-attributed graph scenario, if enriched with appropriate node positional features, can be used to process the topology of the input graph with the purpose of mapping the obtained node representation in a 2D layout minimizing several provided loss functions. We built on the Neural Aesthete component of the framework, to introduce the concept of *fake nodes*, the linear splines control points, that can be optimized via the same underlying optimization process of the graph layout generation. Our contribution can be injected into heterogeneous optimization schemes, in principle involving multiple criteria at once. The idea of handling graph arcs as curves can be found in the very recent work by Yu et al. [27], where a reformulation of gradient descent based on a Sobolev-Slobodeckij inner product enables rapid progress toward local minima of loss functions.

3   Method

We denote a graph as G = (V, E), where $V = \{v_1, ..., v_n\}$ is a finite set of N nodes and $E \subseteq V \times V$ is the edge set representing the arcs connecting the nodes. We further represent each vertex $v_i \in V$ with coordinates $p_i : \mathbb{V} \longmapsto \mathbb{R}^2$, mapping the ith node to a bi-dimensional space. The node coordinate matrix is denoted as $\mathbf{P} \in \mathbb{R}^{N \times 2}$. Finally, the neighborhood of node ith is denoted by N_i.

In literature, a number of graph drawing algorithms have been devised, optimizing functions that express aesthetic index with the advantage of graph topology, geometry, visualization and theory concepts in bi-dimensional or tri-dimensional space [5]. The most typical aesthetic criteria take into account the uniformity of vertex allocation [13], the degree of edge intersections, [19], or the angles between crossing or adjacent edges. It is commonly assumed that graph drawing consists of finding the best vertex allocation when given the adjacency matrix. The latter drives the arcs drawing by connecting linked vertices with segments. In this paper, however, we propose to employ non-uniform splines instead of segments to enhance the readability of the graph. For the sake of simplicity, we restrict our work to linear splines computed by means of *fake vertices* (control points). This allows to improve the readability of graph drawn when employing different GD-based methods, such as the GND or the Stress optimization. More precisely, we propose to introduce splines whenever the GD algorithm is not capable of accomplishing a certain aesthetic criterion while just employing segments. The ratio behind this idea is that the higher degree of freedom available to the optimization algorithm should allow achieving the required aesthetic criterion. The latter is measured by means of the Neural Aesthete, a neural network based model trained to measure the fulfillment of any (even non-differentiable) aesthetic criterion, as explained in Sect. 3.1. In Sect. 3.2, we better clarify how splines are introduced. In Sect. 3.3 we show how the same Neural Aethete provides a loss function that can be employed for Graph Drawing (as shown [21]). Finally, in Sect. 3.4 we show how we can also pair it with other standard GD-based methods, such as Stress optimization.

3.1 Learning Non-differentiable Aesthetic Criteria: The Neural Aesthete

As previously proposed in [21], we employ here the Neural Aesthete, a neural network-based model, capable of learning - and therefore measuring - any aesthetic criteria. Standard GD methods, indeed, are limited by the requirement of employing a differentiable aesthetic criterion to draw a graph. On the contrary, the Neural Aesthete is capable of learning also non-differentiable criteria such as the edge intersections. As done in [21] and without loss of generality, in this paper we train a neural network to measure edge intersection.

To train the Neural Aesthete model, we provide in input a couple of arcs $x = (e_u, e_v)$ and the model return the probability that the arcs are intersecting or not. Each arc is in turn defined as the couple of vertex coordinates $e_u = (p_i, p_j)$, with $e_u \in \mathbf{E}$ and $p_i \in [0, 1]^2$. The model $\nu(\cdot, \cdot, \cdot) : \mathbf{E}^2 \times \mathbb{R}^m \to R$ is working on the two arcs e_u and e_v and returns the output as

$$y_{e_u, e_v} = \nu(\theta, e_u, e_v) \tag{1}$$

where $\theta \in \mathbb{R}^m$ is the vector which represents the weights of the neural network. The learning of the model is based on optimization of a cross-entropy loss function $L(y_{e_u, e_v}, \hat{y}_{e_u, e_v})$, which is defined as:

$$L(y_{e_u, e_v}, \hat{y}_{e_u, e_v}) = -(\hat{y}_{e_u, e_v} * \log(y_{e_u, e_v}) + (1 - \hat{y}_{e_u, e_v}) * \log(1 - y_{e_u, e_v})), \tag{2}$$

where y_{e_u, e_v} is the target label:

$$y_{e_u, e_v} = \begin{cases} 1, & \text{if } (e_u, e_v) \text{ do intersect} \\ 0, & \text{otherwise} \end{cases} \tag{3}$$

Notice that the label y is computed by solving the system of equations of the lines passing by the vertices of the arcs. The solution of this system, if exists, however, is either 0 or 1, therefore it cannot be directly optimized through gradient descent. An artificial dataset composed of 100k input-target entries (x, y) is employed for training the Neural Aesthete. The coordinates $p_{i,j}$ of the vertices of the input arcs are randomly chosen in the interval $[0, 1]^2$. The dataset is balanced to have samples for both classes (cross/no cross). The model is implemented as a MultiLayer Perceptron (MLP) composed of 2 hidden layers, each of 100 nodes with ReLu activation functions. The model is trained to minimize the cross entropy loss in terms of target values, taking maximum advantage through Adam optimizer. A dataset of 50k entries is taken to test the generalization capabilities of the model and achieved an accuracy of 97%.

At test time, the learning process yields a model capable to calculate the probability of intersection between any 2 given arcs. The Neural Aesthete is therefore capable to provide a differential smooth function that guides the positioning of node coordinates via gradient descent. More in details, the provided loss function is the cross-entropy towards respecting the given aesthetic critera - i.e., in this case, towards non-intersection. The optimization parameters, however, are not anymore the weights of the neural networks, but the coordinates of the vertices of the intersecting arcs $(e_u, e_v) = (p_i, p_j), (p_h, p_k)$ in input to the model.

3.2   Employing Splines to Improve Graph Readability

The major contribution of this research is the introduction of splines to connect the arcs which are predicted to not satisfy a certain aesthetic criterion. To predict the satisfaction degree, the previously trained Neural Aesthete model predicting edge intersection is employed. To introduce splines, we add a *fake vertex* on one of the intersecting arcs. Consequently, the arc will be optimized by employing more nodes instead of only the two extremities, as it is split and considered as two separated arcs. At each iteration, a batch of random arcs are chosen as input and the trained Neural Aesthete outputs the degree of intersection of the given arcs. Whenever two arcs are predicted as intersecting, a fake node is generated to further help the optimization process and reduce the probability of intersection. Since we want to create a spline, we add two arcs to the adjacency matrix connecting the new node to the previous ones, and we remove the old arc. The newly introduced spline will help the optimization process and by enhancing the freedom of the layout generation process, resulting in more optimized results. The Graph Drawing process is then conducted on all the vertices. For the sake of simplicity, and to avoid degenerated solutions, we avoid introducing splines during the first T iterations. Also, we limit to S_{max} the number of splits for each starting arc in the input graph.

Going into more details, the Neural Aesthete v process a random arc pair (e_u, e_v) picked from the arc list E, to predict their degree of intersection $\hat{y}_{e_u, e_v} = \nu(\theta, e_u, e_v)$. If the two arcs are intersecting each other, the first arc is split with the creation of a *fake vertex* $p_f = \frac{p_i + p_j}{2}$. The arc list E is updated, and the node coordinates of the created vertex are added to the parameters of the optimization process. In case the arcs created when introducing the spline are randomly chosen, the coordinates of the new vertex is also processed. This is due to the fact that we want to further optimize the spline. This helps in optimizing edges while using more points and possibly by further splitting the arcs.

3.3   Edge Crossing Optimization with Splines

As previously introduced, the same Neural Aesthete can be employed in a gradient descent-based optimization process to display an input graph. The employed loss function $L(\cdot, \cdot)$ is the cross entropy loss introduced in Eq. 2 computed on the randomly chosen arcs e_u, e_v with respect to the non-intersection $y_{e_u//e_v} = 0$.

$$H_{u,v} = L(\hat{y}_{e_u,e_v}, y_{e_u//e_v}) = -\log(1 - \hat{y}_{e_u,e_v}) \tag{4}$$

This smooth and differential loss function increases the utilization of gradient descent methods to optimize edge coordinates (e_u, e_v) with respect to any learned aesthetic criterion. This is process is replicated for all graph edges,

$$H(\mathbf{P}) = \sum_{(e_u,e_v)\in\mathbf{E}} L(\hat{y}_{e_u,e_v}, y_{e_u//e_v}) \tag{5}$$

A possible scheme for graph drawing is then:

$$\mathbf{P}^* = \arg\min_{\mathbf{P}} H(\mathbf{P}) \tag{6}$$

A feasible solution to carry out this is by using optimization methods as gradient descent:

$$\mathbf{P} \longleftarrow \mathbf{P} - \eta \, \nabla_{\mathbf{P}} H(\mathbf{P}) \tag{7}$$

where η represents the learning rate.

The remarkable progress given by this framework is the computational efficiency and parallelization capabilities of neural networks, since the prediction of edge crossing can be carried out for multiple arc pairs in parallel. However, the optimization process is greatly facilitated when introducing splines, since both new arcs are moved in the directions where they are no more intersecting, participating in reducing the number of cross intersections and improving the graph layouts.

3.4   Stress Optimization with Splines

The Stress function [?], is one of the empirically proved techniques to be of great importance in drawing aesthetically pleasing graph layouts through pleasing node coordinate selection. The optimized loss function is defined as:

$$STRESS(P) = \sum_{n=1} w_{ij}(\| p_i - p_j \| - d_{ij})^2, \tag{8}$$

where $p_i, p_j \in [0;1]^2$ are the coordinates of vertices i and j respectively. The graph theoretic distance d_{ij} is the shortest path between node i and j, and w_{ij}

is a normalization factor balancing the impact of the pairs, and it is defined as $w_{ij} = d_{ij}^{-\alpha}$ with $\alpha \in [0, 1, 2]$[1]. The graph drawing capabilities of stress optimization allows generating pleasing layouts even when employing segments to connect vertices. However, in this work, we also tested the employment of splines along with stress optimization. The steps exposed in Sect. 3.2 to generate splines are performed in this case as well.

We perform stress optimization on the vertices coordinates by minimizing the stress function (see Eq. 8). For each graph, we compute the shortest path d_{ij} among every node pair (i, j) and the loss is calculated based on the shortest path and the distance calculated for each node pair. The graph layouts clearly presents pleasing layouts as a result of the stress optimization. Further, a decreased number of cross intersection of arcs is observed in the graphs after splines are introduced in the model.

4   Experiments

We devised two different experimental settings to prove the effectiveness of our proposal. We compare the graph layouts produced by different combinations of loss functions, optimized by Gradient Descent. We implemented our method exploiting the PyTorch framework and we used the NetworkX [14] python package for the graph visualization utilities.

The main goal of the first experiment is the reduction of the number of arc intersections by injecting our spline-based method into an optimization process guided by the Neural Aesthete. Hence, the Neural Aesthete is exploited both to identify the crossing edges and to move the node coordinates (both the real and *fake* vertices, once they are introduced) towards the direction of non-intersection. In order to prove the efficiency of our method, we tested our approach over 9 heterogeneous graph classes (see Table 1), considering graphs with different characteristics and properties. Overall, we focused on graphs having small sizes, given that previous works highlighted how node-link layouts and cross-intersection based losses are more suitable to graphs with small size [25].

We compare the number of arc intersections produced, at the end of the layout optimization process, by optimizing solely the Edge-crossing (EC) loss provided by the Neural Aesthete, with the case in which we combine this optimization process with the proposed spline-based method (EC+SPLINES). We carried on the optimization via mini batches composed by 2 arc-couples, for an amount of $10k$ iterations (gradient steps), with a learning rate $\eta = 10^{-2}$. The initial node coordinates are randomly picked in the interval $(0, 1)$. We provide in Table 1 a quantitative comparison of these two approaches in the considered graph classes.

---

[1] We tested our work with stress optimization of graphs using different weighing factor α. However, the best graph layouts were empirically observed with $\alpha = 2$.

Table 1. Cross intersection reduction. We tested the proposed approach in 9 different graph classes (first column). We report the number of arc intersections at the end of the drawing optimization process, when minimizing the edge-crossing loss from the Neural Aesthete only (EC column), and when combining it with the proposed spline-based method (EC+SPLINES column).

Graph Class	EC	EC+SPLINES
Karate	142	118
Circular	25	18
Cube	3	0
Tree	3	2
Diamond	0	0
Bull	0	0
Simple	1	1
Cycle	1	1
Barbell	22	22

Thanks to the adoption of the proposed method, the number of intersection is reduced in most of the considered settings. We remark that the introduction of *fake* vertices and the corresponding arcs hinders, in principle, the optimization process, given that the amount of edges and arc is increased. Nevertheless, in most of the cases, the Neural Aesthete is capable to produce improved layouts at the end of the learning process. We show in Fig. 2 qualitative examples of the produced layouts. First column depicts the node positions produced by the standard EC loss optimization. When injecting the EC+SPLINES loss, the number of intersections is reduced thanks to inherited improved freedom in node position selection.

To further show the general nature of our proposal, we injected the proposed *fake* vertices creation into an optimization process minimizing the stress loss. We tested this method over the same graph classes of the previous experimental setup, with mini batches having the size of half the arc numbers of the input graph. We optimized the loss for an amount of $10k$ iterations (gradient steps). We set the learning rate $\eta = 10^{-1}$. We report in the third and fourth column of Fig. 2 the layouts obtained via optimizing the stress loss solely (STRESS column) and combining the stress loss with the proposed spline-based method (STRESS+SPLINES column). We can see how the obtained graphs still hold the visual characteristics specfi to the stress minimization, while reducing the number of intersecting arcs (evident in Diamond, Circular ladder).

E.C.	E.C. + Splines	Stress	Stress+ Splines

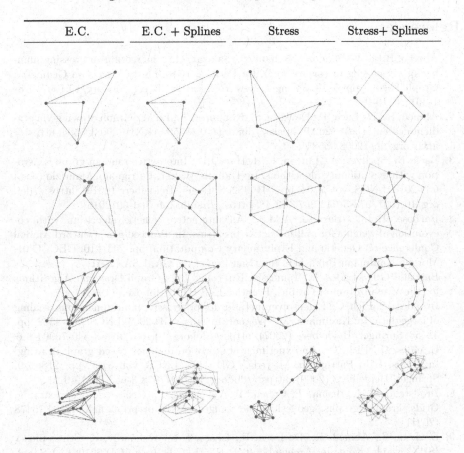

Fig. 2. Qualitative analysis on the obtained graph layouts. Different graph classes (rows: Diamond, Cube, Circular-Ladder and Barbell graphs) are plot using the 4 optimization paradigms (columns) investigated in the experimental experience.

5   Conclusion

We proposed a framework that improves graph layouts leveraging the tool of linear splines. We exploit a Neural Aesthete to devise if arcs are crossing, and if it is the case, we split them into spline-based segments. We show how this approach can be injected into a common Graph Drawing scenario based on gradient Descent optimization, both when optimizing the number of arcs intersections and common aesthetic losses such as the stress function. Future work will be focused on extending the proposed solution into Graph Neural Drawers powered by Graph Neural Networks, in order to speed up the graph drawing pipeline and the optimization processing time.

References

1. Ábrego, B.M., Fernández-Merchant, S., Salazar, G.: The rectilinear crossing number of k_n: closing in (or are we?). In: Pach, J. (ed.) Thirty Essays on Geometric Graph Theory, pp. 5–18. Springer, New York (2013). https://doi.org/10.1007/978-1-4614-0110-0_2

2. Ahmed, R., De Luca, F., Devkota, S., Kobourov, S., Li, M.: Graph drawing via gradient descent, $(gd)^2$ (2020). https://doi.org/10.48550/ARXIV.2008.05584, https://arxiv.org/abs/2008.05584

3. Bach, B., Spritzer, A., Lutton, E., Fekete, J.D.: Interactive random graph generation with evolutionary algorithms. In: Didimo, Walter, Patrignani, Maurizio (eds.) GD 2012. LNCS, vol. 7704, pp. 541–552. Springer, Heidelberg (2013). https://doi.org/10.1007/978-3-642-36763-2_48, https://hal.inria.fr/hal-00720161

4. Barbosa, H.J.C., Barreto, A.M.S.: An interactive genetic algorithm with co-evolution of weights for multiobjective problems. In: Proceedings of the 3rd Annual Conference on Genetic and Evolutionary Computation, pp. 203–210. GECCO'01, Morgan Kaufmann Publishers Inc., San Francisco, CA, USA (2001)

5. Battista, G.D., Eades, P., Tamassia, R., Tollis, I.G.: Graph Drawing: Algorithms for the Visualization of Graphs. Prentice Hall PTR, Hoboken (1998)

6. Brandes, U., Pich, C.: Eigensolver methods for progressive multidimensional scaling of large data. In: Kaufmann, M., Wagner, D. (eds.) GD 2006. LNCS, vol. 4372, pp. 42–53. Springer, Heidelberg (2007). https://doi.org/10.1007/978-3-540-70904-6_6

7. Brandes, U., Pich, C.: An experimental study on distance-based graph drawing. In: Tollis, I.G., Patrignani, M. (eds.) GD 2008. LNCS, vol. 5417, pp. 218–229. Springer, Heidelberg (2009). https://doi.org/10.1007/978-3-642-00219-9_21

8. Bronstein, M.M., Bruna, J., Cohen, T., Veličković, P.: Geometric deep learning: Grids, groups, graphs, geodesics, and gauges. arXiv preprint. arXiv:2104.13478 (2021)

9. Devkota, S., Ahmed, R., De Luca, F., Isaacs, K.E., Kobourov, S.: Stress-plus-x (SPX) graph layout. In: Archambault, D., Tóth, C.D. (eds.) GD 2019. LNCS, vol. 11904, pp. 291–304. Springer, Cham (2019). https://doi.org/10.1007/978-3-030-35802-0_23

10. Didimo, W., Liotta, G., Montecchiani, F.: A survey on graph drawing beyond planarity. ACM Comput. Surv. (CSUR) **52**(1), 1–37 (2019)

11. Eades, P., Hong, S.-H., Klein, K., Nguyen, A.: Shape-based quality metrics for large graph visualization. In: Di Giacomo, E., Lubiw, A. (eds.) GD 2015. LNCS, vol. 9411, pp. 502–514. Springer, Cham (2015). https://doi.org/10.1007/978-3-319-27261-0_41

12. Frick, A., Ludwig, A., Mehldau, H.: A fast adaptive layout algorithm for undirected graphs (extended abstract and system demonstration). In: Tamassia, R., Tollis, I.G. (eds.) GD 1994. LNCS, vol. 894, pp. 388–403. Springer, Heidelberg (1995). https://doi.org/10.1007/3-540-58950-3_393

13. Gibson, H., Faith, J., Vickers, P.: A survey of two-dimensional graph layout techniques for information visualisation. Inf. Vis. **12**(3–4), 324–357 (2013)

14. Hagberg, A., Swart, P., S Chult, D.: Exploring network structure, dynamics, and function using networkx. Technical report, Los Alamos National Lab. (LANL), Los Alamos, NM (United States) (2008)

15. Jacomy, M., Venturini, T., Heymann, S., Bastian, M.: Forceatlas2, a continuous graph layout algorithm for handy network visualization designed for the gephi software. PLoS ONE **9**(6), e98679 (2014)

16. Kamada, T., Kawai, S., et al.: An algorithm for drawing general undirected graphs. Inf. Process. Lett. **31**(1), 7–15 (1989)
17. Kwon, O.H., Ma, K.L.: A deep generative model for graph layout. IEEE Trans. Visual Comput. Graphics **26**(1), 665–675 (2019)
18. Paszke, A., et al.: Automatic differentiation in pytorch (2017)
19. Purchase, H.: Which aesthetic has the greatest effect on human understanding? In: DiBattista, G. (ed.) GD 1997. LNCS, vol. 1353, pp. 248–261. Springer, Heidelberg (1997). https://doi.org/10.1007/3-540-63938-1_67
20. Scarselli, F., Gori, M., Tsoi, A.C., Hagenbuchner, M., Monfardini, G.: The graph neural network model. IEEE Trans. Neural Netw. **20**(1), 61–80 (2009)
21. Tiezzi, M., Ciravegna, G., Gori, M.: Graph neural networks for graph drawing. In: IEEE Transactions on Neural Networks and Learning Systems, pp. 1–14 (2022). https://doi.org/10.1109/TNNLS.2022.3184967
22. Tiezzi, M., Marra, G., Melacci, S., Maggini, M.: Deep constraint-based propagation in graph neural networks. In: IEEE Transactions on Pattern Analysis and Machine Intelligence (2021)
23. Tiezzi, M., Marra, G., Melacci, S., Maggini, M., Gori, M.: A lagrangian approach to information propagation in graph neural networks. In: ECAI 2020–24th European Conference on Artificial Intelligence. Frontiers in Artificial Intelligence and Applications, vol. 325, pp. 1539–1546. IOS Press (2020). https://doi.org/10.3233/FAIA200262
24. Wang, X., Yen, K., Hu, Y., Shen, H.: Deepgd: a deep learning framework for graph drawing using GNN. CoRR abs/2106.15347 (2021), https://arxiv.org/abs/2106.15347
25. Wang, Y., Jin, Z., Wang, Q., Cui, W., Ma, T., Qu, H.: Deepdrawing: a deep learning approach to graph drawing. IEEE Trans. Visual Comput. Graph. **26**(1), 676–686 (2019)
26. Wang, Y., et al.: Revisiting stress majorization as a unified framework for interactive constrained graph visualization. IEEE Trans. Visual Comput. Graph. **24**(1), 489–499 (2017)
27. Yu, C., Schumacher, H., Crane, K.: Repulsive curves. ACM Trans. Graph. (TOG) **40**(2), 1–21 (2021)
28. Zheng, J.X., Goodman, D.F.M., Pawar, S.: Graph drawing by weighted constraint relaxation. CoRR abs/1710.04626 (2017), http://arxiv.org/abs/1710.04626
29. Zheng, J.X., Pawar, S., Goodman, D.F.: Graph drawing by stochastic gradient descent. IEEE Trans. Visual Comput. Graph. **25**(9), 2738–2748 (2018)

Multi-stage Bias Mitigation for Individual Fairness in Algorithmic Decisions

Adinath Ghadage[1], Dewei Yi[1], George Coghill[1], and Wei Pang[2(✉)]

[1] University of Aberdeen, Aberdeen, UK
{A.Ghadage.18,Dewei.Yi,George.Coghill}@abdn.ac.uk
[2] Heriot-Watt University, Edinburgh, UK
W.Pang@hw.ac.uk

Abstract. The widespread use of machine learning algorithms in data-driven decision-making systems has become increasingly popular. Recent studies have raised concerns that this increasing popularity has exacerbated issues of unfairness and discrimination toward individuals. Researchers in this field have proposed a wide variety of fairness-enhanced classifiers and fairness matrices to address these issues, but very few fairness techniques have been translated into the real-world practice of data-driven decisions. This work focuses on individual fairness, where similar individuals need to be treated similarly based on the similarity of tasks. In this paper, we propose a novel model of individual fairness that transforms features into high-level representations that conform to the individual fairness and accuracy of the learning algorithms. The proposed model produces equally deserving pairs of individuals who are distinguished from other pairs in the records by data-driven similarity measures between each individual in the transformed data. Such a design identifies the bias and mitigates it at the data preprocessing stage of the machine learning pipeline to ensure individual fairness. Our method is evaluated on three real-world datasets to demonstrate its effectiveness: the credit card approval dataset, the adult census dataset, and the recidivism dataset.

Keywords: Algorithmic bias · Algorithmic fairness · Fairness-aware machine learning · Fairness in machine learning · Individual fairness

1 Introduction

With the widespread use of machine learning algorithms in decision-making systems, concerns about trust in AI are growing in terms of its full adaptation. The decisions in many decision-making systems are based upon the predictions of the results of machine learning algorithms. The major challenge for policy-makers, stakeholders, and companies in adopting AI is the black box nature of AI-based decision-making systems [31]. Recently, some studies [23,34] attempted to open the black box of AI. The researchers have proposed new fundamental pillars for

© The Author(s), under exclusive license to Springer Nature Switzerland AG 2023
N. El Gayar et al. (Eds.): ANNPR 2022, LNAI 13739, pp. 40–52, 2023.
https://doi.org/10.1007/978-3-031-20650-4_4

trust in the AI system, including explainability, fairness, robustness, and lineage [3,4,26].

Fairness is one of the fundamental principles of a trustworthy AI system [32]. Systematically understanding bias specifically against each individual as well as group members in the dataset defined by their protected attributes like age, gender, race, and nationality is the first step to achieving fairness in decision-making and building trust in AI in general. The protected attributes, concerned in this paper follow the fairness guidelines given by the Information Commissioner's Office (ICO) in terms of Equality Law 2010 in the UK. Among these guidelines, ICO proposes potentially protected attributes including age, disability, gender, marital status, maternity, race, religion, sex, and sexual orientation. According to a recent study, machine learning algorithms treat people or groups of people who have the above-mentioned protected attributes unfairly. Our research attempts to identify potential research gaps between existing fairness approaches and possible techniques to address the fair classification of machine learning algorithmic decision-making systems to contribute to building trust in AI. Many studies have been done on group fairness (which is also called statistical parity). This family of definitions fixes a small number of protected groups, such as gender, race, and then approximate parity of some statistical measure across all of these groups. Some of the popular measures include the false positive rate and the false negative rate, which are also known as equalise odd and equality of opportunity [8,9,13,18,19,21,33], respectively, but fewer are concerned with individual fairness. The group fairness approaches (e.g. statistical parity, equal opportunity, and disparate mistreatment) are used to investigate discrimination against members of protected attributes such as age, gender, and race. A fair classifier tries to achieve equality across the protected groups like statistical parity [13], equalised false positive and false negative rates, and calibrations [6].

Most of the group fairness definitions are subjected to learning statistical constraints [10] or averaged over the protected groups to satisfy fairness definitions [10,16]. As group fairness is measured by aggregating over male or female or any other protected attribute, this constraint-based definition may harm some of the individuals within that group. Individual fairness is an alternative approach that satisfies the constraints for specific pairs of individuals defined by their task similarity. The notion of individual fairness is defined by *"similar individuals should be treated similarly."* [13]. Here, similarity is defined in terms of task-specific similarity metrics, where a classifier maps individuals to the probability distribution of outcomes. For example, if x_i and x_j are similar to each other, then their classification predictions y_i and y_j need to be the same.

In this paper, we address the individual notion of fairness based on the work done in [1,13,22,34], in the sense that we attempt to understand more fundamental questions about how an individual is classified as fair/unfair in task-specific similarity. A model is identified to be fair to individuals if similar pairs of individuals yield similar outcomes in prediction. The similarity of individuals is determined by the closeness of distance between the data points in the input space, which satisfies the Lipschitz property (i.e., distance preservation). The model is unfair to individuals if similar individuals are treated discriminatorily in their predictions. That is, for two similar individuals $< x_i, x_j >$, their classification predictions are different, that is, $y_i \neq y_j$. To estimate a model's individual unfairness, we can use a pool of similar individuals generated by a human specified process from the original data and/or from the transformed data and their discrimination in treatment among these individuals. In our proposed approach, we revisit the notion of individual fairness proposed by Dwork et al. [13], that is, similar individuals are treated similarly on the same given task. We learn to generalise a representation of the original data into a transformed representation. The transformed representation learnt by our model preserves fairness awareness similarity among data points with multiple protected attributes considered rather than the single protected attribute used in much of the previous individual fairness research work [30]. In this research, the words "sensitive" and "protected" are used interchangeably for the same purpose to specify the list of possible attributes based on the individuals who can be treated discriminatorily in their predictions.

Furthermore, in our work, we aim to identify and mitigate the bias presented in the data by historical decisions. The pre-processing approach is enforced to reduce discrimination and make the model fairer to individuals. We applied pre-processing techniques on transformed data to identify and remove biased data points. The process of removing biased data in our method is to modify those outcome labels of similar pairs of individuals, where such pairs contribute more to the model's unfairness and leave all other data points and features unchanged. Modifying the outcome values of similar pairs yields a less biased dataset. A model is then trained on these less biased samples and produces a fairer outcome with less individual discrimination than the model trained on the original data. Our fairness model offers more adaptability and versatility to data with multiple sensitive attributes. The experiments performed on three real-world datasets with only individual fairness definitions are considered, excluding the group fairness definitions. More specifically, the contributions of this paper are summarised as follows:

- We propose a novel approach to improve individual fairness. To the best of our knowledge, this is the first attempt to provide individual fairness by considering multiple sensitive attributes to identify and mitigate biases in the dataset.
- We develop an application-agnostic feature transformation approach by learning transformed representations of data points that restore individually fair data and accuracy such that application-specific multi-valued multiple sensitive attributes are considered, rather than single binary protected attributes such as gender.
- Our method can identify and mitigate bias at the pre-processing stage of the machine learning pipeline.
- Our method is evaluated on classification and regression tasks, showing that strong individual fairness can indeed reconcile with a high utility on real-

world datasets: the adult census, credit card approval dataset and recidivism dataset.

2 . Background and Related Work

2.1 Statistical Definitions of Fairness

Most of the research on fairness attempts to deal with two missions: 1) developing methods to detect bias and discrimination in AI-based decision-making systems and 2) developing methods to mitigate these biases by using different criteria to improve fairness in AI-based systems. There are a wide variety of bias metrics and fairness definitions proposed in the literature [10,14,17,23,24,31]. The group fairness [13] is a constraint-based approximation of parity across all groups with a statistical measure. Suppose a and a' are the classes of protected and unprotected groups, and the group fairness constraint to satisfy equal probability of prediction across these two groups is defined as $P[Y|A = a] = P[Y|A = a']$. Here, bias metrics quantify system error in the context of fairness and bias systematically, which provides advantages to privileged groups over disadvantages to unprivileged groups. Bias mitigation algorithms reduce unwanted bias in the data. Conditional parity [11,15,25,31] and the inequality indices [5,28,29] are the most commonly used statistical measures of fairness, where this definition satisfies the equal prediction of outcome in both protected and unprotected groups controlled by legitimate factors. However, it is impossible to implement all these definitions in practice as there is no guideline on which bias metrics and bias mitigation algorithms should be used to address any specific definition of fairness. Therefore, despite recent awareness of bias and fairness issues in AI development and deployment, there is no systematic operation in practice.

2.2 Definitions of Individual Fairness

Alternatively, the individual notion of fairness is a constraint over pairs of individuals rather than an average over group members. Individual fairness ensures that members who are similar are treated similarly. Here, the similarity is a task-specific similarity metric that must determine the basis of this notion of definitions. Assume that metrics-based fairness defines similarity on variables (i.e., input feature vectors) as follows: $m : V \times V \to \mathbb{R}$ m is a map function which maps each individual to distribution of outcomes. Hence, metric fairness for individuals with variable $v, v' \in V$ is a closeness in their decisions.

$$|f(v) - f(v')| \leq m(v, v') \tag{1}$$

This formulation is based on Lipschitz condition [13,27].

The definitions of metrics in individual fairness change subsequently, authors in [13] define the task-specific similarity metrics over *individuals*. In the following research, metrics are defined over features, variables, and inputs to classifiers [27]. A construction space was introduced in [16] in addition to the observed space

(OS) and decision space (DS) [22]. A construction space (CS) is a metric space consisting of individuals and their distances. Whereas, Observed space (OS) is a metric space which approximates metrics in CS with respect to task, assuming $g : v \rightarrow v'$ that generates an entity $v' = g(v)$ from a person $v \in CS$.

Another notion of individual fairness is proposed in [27] where the authors approximate the metrics of fairness by marginalising a small probability of error in the similarity between two individuals. This is an extension of metrics fairness defined by Dwork et al. [13]. In this matrix approximate fairness [27] definition two constants α, γ are used to approximate in addition to the similarity metrics definition suggested by Dwork [13]. In AI Fairness360 [4], it implements individual fairness mapping as the author proposed methods that measure similar prediction of a given instance to its nearest neighbours [34]. Similarly, in average individual fairness, [20] method is inspired by oracle-efficient algorithms.

3   Multi-stage Individual Fairness

In this section, we explain the whole framework of our proposed method to investigate unfairness in machine learning framework. After having a comprehensive literature review, our research contributions are mentioned in Sect. 1. To address these research contributions, several novel techniques are proposed in this research. The rest of this section describes the details of each stage and its responsibilities (Fig. 1).

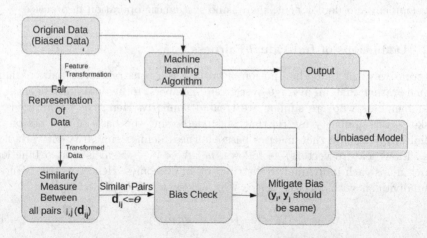

Fig. 1. Flowchart with step in our individual fairness model.

We consider that a fair algorithm should consider both protected and non-protected attributes while making fair decisions, and existing approaches give less attention to the multiple protected attributes in both group and individual fairness. Our work will be a further step towards individual fairness [13], fair [22] and perfect metrics fairness [27] with a focus on multiple protected attributes. These methods measure similarity by Lipschits mapping between a pair of individuals in unknown distributions over the classified outcomes with distance metrics. Perfect metrics fairness [27] is a generalised approach that approximates individuals' lying in both group and individual fairness. Here, Lipschits' mapping conditions map task-specific similarity metrics. Our proposed framework for individual fairness consists of detecting and mitigating bias at the preprocessing stage of the machine learning pipeline. The term "IndFair" is referred to as our defined approach to individual fairness in the paper. The proposed framework consists of five major components in the machine learning fairness pipeline. These components are as follows: fair representation of data, similarity measure, bias identification, mitigating identified bias, and the final unbiased model output. In the first component, a fair representation of features is computed using the transformation function detailed in this section. In the second component, we measure the similarity between a pair of individuals by using the Euclidean distance function. We identify a bias in the similarity measured data in the third component. Fairness measures and bias mitigation are performed in the preprocessing stage of the machine learning pipeline. A detailed description of the working principle behind each component is given in the rest of this section.

3.1 Notations

We define the notations used in the proposed model in which the input data X is represented as a $m \times n$ matrix where each individual in the populations as $Xi = 1, 2, 3, ..., n$ and m is the number of features. Each person x_i has m features (i.e. variables, input) $x_i \in X$, where features m is a combination of protected features p and non-protected features np. We assume the attributes $1...l$ protected attributes and the attributes $l+1...m$ are non-protected. A binary classification decision on each person is denoted as $\hat{Y} = f(X, Y)$, where f is a function of variables known at decision time $f : X \longrightarrow 0, 1$. In binary prediction based system a outcome variable (i.e. predictor) $\hat{Y}$ for each person is unknown at decision time and the actual outcome is denoted by $Y = (y_1, y_2, ..., y_n)$.

3.2 Transformed Representation Learning

We aim to transform features into fair representation by matrix multiplication in the transform stage. Here, the intuition of matrix multiplication is vectors of protected and non-protected attributes are multiplied by distributive property. Each person x_i has m features (i.e. variables, input) $x_i \in X$, where features m is a combination of protected features p and non-protected features np. We assume the attributes $1...l$ protected attributes and the attributes $l + 1...m$ are non-protected. Each x_p is the vector representation of the protected attributes for individual i, similarly x_{np} is vector of non-protected attributes. The result of this transformation can be viewed as a low-level representation of individual i with $k = m - l - 2$ dimensions vector size of attributes. We perform the above operation for all data points. The mapping of $x_i \rightarrow \tilde{x_{i_k}}$ is given below:

$$\tilde{x_{i_k}} = \sum_{p=1}^{l} x_p \left(\sum_{np=l+1}^{m} x_{np} \right) \tag{2}$$

3.3 Similarity Measure

The similarity measure is an important component to achieve individual fairness in algorithmic decision making in the pre-processing stage. Similarity measure can be achieved through a distance measure between two individual records in the distribution of feature space. Mostly used distance functions are Euclidean distance, Manhattan distance, and Minkowski distance. In this paper, we focus on the euclidean distance function to measure the similarity between all pairs of individuals.

$$d(x_i, x_j) = \sqrt{\sum_{i,j=1}^{n} (x_i - x_j)^2} \tag{3}$$

The above distance functions d presented in Eq. (3) is applicable to original records x_i and transformed records $\tilde{x_{i_k}}$ to measure the similarity among all pair of records.

3.4 Fairness Measure

Mapping Individual Bias: The individual fairness is to preserve the fairness-aware distances between a pair of individuals i and j in a given matrix space. The mapping of individual bias in the given matrix space is measured as the *consistency* of outcome between the pairwise similar individuals in transformed data and original data. The matrices used to measure the individual bias captures the intuition of individual similarity definition, that is, the similar individuals should be treated similarly. For instance, if two individual records x_i and x_j are similar, then we check the consistency of the output variable y_i and $\tilde{y_i}$. We adapted the bias mapping metrics defined in iFair [22] and [34], where our formula is different from the one used in iFair. The distance d is on a paired records x_i, x_j of input data X, and the transformed data $\tilde{X}$ with each pair of the records $\tilde{x_i}, \tilde{x_j}$ whereas $\tilde{d}$ is pairwise distance on transformed records. The mapping of bias in input data and transformed data is performed using fairness loss F_{loss} is as given below,

$$F_{loss}(X, \tilde{X}) = 1 - \sum_{i,j=1}^{n} |(d(x_i, x_j) - \tilde{d}(\tilde{x_i} - \tilde{x_j})|, \tag{4}$$

Mitigating Bias: The bias mapping seeks to identify any unfair distortion in the original data, transformed data and output data for mitigating the bias in the pre-processing stage. The Eq. 4 a fairness loss that is, F_{loss} which addresses a *systematic or structural bias* presented in the data. The intuition to mitigate bias in fair individual classifications is based on the similarity of the outcome of two individuals i and j. If the similarity distance function $d(x_i, x_j)$ indicates that individual records, i and j are similar in transformed data then their outcome value y_i and y_j should be similar. If these individuals outcome is not similar then we modify their outcome value to mitigate bias using the below formula.

$$Y_i, Y_j = \begin{cases} 1, & \frac{1}{n}\left((y_i, y_j = 1)\right) \geq 0.5; \\ 0, & \text{otherwise} \end{cases} \tag{5}$$

In the above, y_i, y_j are the mitigated values for similar individuals. Here, the intuition is for similar pair of individuals we take an average value of binary outcome either 0 or 1. If this average values of outcome for individuals holding positive outcome e.g., $\frac{1}{n}y_i, y_j = 1 \geq 0.5$ then the outcome of all these similar individuals is replaced as 1 and 0 otherwise. We used the same bias mitigation strategy in the $pre-processing$ stage of fairness pipeline defined in [4].

3.5 Optimisation

Utility: The utility measured as accuracy (Acc) for each of the support vector machine classifiers (SVM) and logistic regression(LR) for the classification task. We adapted the utility measure from iFair [22], as given below:

$$Utility(Y, \tilde{Y}) = \sum_{i=1}^{n}(y_i - \tilde{y}_i)^2 \tag{6}$$

Overall Objective Function: Combining the data loss i.e., utility 6 and the fairness loss 4 yields our final objective function, the learned representation is to minimise the objective function. The utility minimises data loss and the individual fairness is fairness-aware treatment in input and output space.:

$$L_{total}(\Theta) = Utility(Y, \tilde{Y}) + F_{loss}(X, \tilde{X}) \tag{7}$$

4 Data and Experiment

4.1 Datasets

We evaluate our method on three real-world datasets, which are publicly available for researcher use. In this experiment, we examine the structural bias in the three different datasets including Adult data [7], Recidivism data [2] and Credit approval dataset [12] for fair individual classification.

– Adult dataset is a census income data in the US [7] which consists of 48,842 records. In this dataset, a target variable Y is individual income with more than 50 thousand dollars and the protected attributes that we used in our framework are *age, marital status, sex and race* for the binary classification task.

– Credit Approval dataset is a collection of 690 records in credit approval application [12]. The binary outcome value Y represents if the individual is default or not. The protected attributes that we used in our experiments are *sex, age, married, ethnicity and citizen*.

– Recidivism dataset is a widely used test case in experiments on fairness algorithms. We have used 11 attributes including 3 protected attributes, namely *age, sex and race*. The target variable two years of recidivism is the binary outcome value.

The data are randomly split into three parts: train set, test set and validation set to learn model parameters and the validation set is used for validation. We use the same setting in our experiments and compare all methods. We train and evaluate the data by using a support vector machine and logistic regression as classifiers and our individual fairness method (IndFair) and learning fair representation [34]. In our setting, the data are used to compute the accuracy and fairness at the pre-processing stage by using the above-mentioned machine learning algorithms and individual fairness definitions. We have only considered support vector machine classifier (SVM) and logistic regression (LR) in our experiments and experiments using neural classifiers will be considered in the future work. These setting of data are given in the Table 1 with attribute name *Method* which contains original data, pre-processed, post-processed, and optimal. The data setup is described in the following section.

Data Setup

– The original data consists of all the attributes including the protected attributes and the non-protected attributes.

– The result of transformed data are *Pre-processed* data given in the Table 1. The data is a transformed representation of original data which preserves the fairness-aware distance between pairs of individuals learned by applying transformed representation learning. We then check the accuracy and fairness of the data.

4.2   Evaluation Measures

– **Utility:** This is measured as accuracy (Acc) on tested classifiers, including the support vector machine (SVM) and logistic regression (LR) where these classifiers work on the binary classification tasks on three different setups of the data as mentioned above.

– **Individual Fairness:** The individual fairness, that is *IndFair*, is measured by the consistency of outcome for the individually fair pairs. This means if the pair of records are similar to each other based on the fact that the distance value is less than the given threshold, then the predicted classification of similarly paired individuals should be the same. We categorise the similar pair individuals into three parts based on their output value y for all the individual records below the threshold distance. The first part of individuals have both positive outcome that is $y_i, y_j = [1, 1]$, in the second part both individuals have a outcome values $y_i, y_j = [0, 0]$, and the third part of individuals have either positive or negative but do not have same outcome value for each pair of individuals such that $y_i, y_j = [1, 0]$ or $[0, 1]$. The mapping of individual fairness is given as follows:

$$IndFair(Y) = 1 - \sum_{i,j=1}^{n} (y_i, y_j) - (\tilde{y}_i \tilde{y}_j) \tag{8}$$

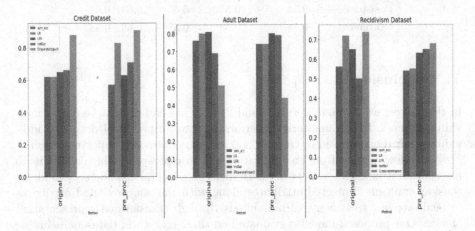

Fig. 2. Experiment result on three datasets with original and pre-processed data setting

4.3 Experimental Results

In this section, we demonstrate the effectiveness of our proposed model on the classification task. The Fig. 2 shows the results for all the methods on three datasets, plotting the accuracy and individual fairness. We observe that there is a considerable amount of unfairness in the original data, therefore the Ind-Fair accuracy is lower in all the datasets. However, the IndFair is significantly increased in pre-processed data. The overall performance of IndFair is better than representation learned by learning fair representation (LFR) in terms of identifying bias and improving the individual fairness in the pre-processing stage of the machine learning pipeline. Table 1 shows the results of the accuracy and fairness trade-off of the machine learning classifier at three categories of data distribution as well as their fairness measure using IndFair and LFR. The optimal results are the harmonic mean of the results in all methods in the datasets setup. The pre-processed method shows an improvement in fairness; however, there is still a considerable size of unfairness hidden in the data.

Table 1. Experimental results for classification and individual Fairness task.

Dataset	Method	SVM ACC.	LR Acc.	IndFair	LFR	DisparateImpact
Adult data	Original data	0.76	0.80	0.69	0.81	0.51
	Pre-processed	0.74	0.74	0.79	0.80	0.44
	Post-processed	0.74	0.74	0.76	0.80	0.65
	Optimal	0.74	0.75	0.74	0.80	0.53
Credit data	Original data	0.62	0.62	0.66	0.65	0.88
	Pre-processed	0.57	0.83	0.71	0.63	0.91
	Post-processed	0.60	0.81	0.69	0.63	0.87
	Optimal	0.59	0.74	0.68	0.63	0.88
Recidivism data	Original data	0.56	0.72	0.50	0.65	0.74
	Pre-processed	0.54	0.55	0.65	0.63	0.68
	Post-processed	0.54	0.55	0.63	0.63	0.62
	Optimal	0.54	0.59	0.58	0.63	0.68

5   Conclusion

In this paper, we propose a generic and flexible method to achieve better individual fairness. It is framework to perform a transformation of data into individually fair representations. Our method accommodates two important criteria. First, we view fairness from an application-agnostic prospect, which allows us to incorporate it in a wide variety of tasks, including general classifiers. Second, we consider multiple protected attributes along with the non-protected attributes to facilitate fair treatments of individuals through transformed representation of data. Our proposed model is evaluated on three real-world datasets including Adult income data, Credit data and Recidivism dataset, which demonstrates that the consistency with utility and individual fairness can reach a promising degree by using our model. With applying the representations of our individual fair model on classifier, it leads that algorithmic decisions through our approach are substantially more fairer than the decisions made on the original data.

References

1. Ahn, Y., Lin, Y.R.: Fairsight: visual analytics for fairness in decision making. IEEE Trans. Visual Comput. Graphics **26**(1), 1086–1095 (2019)
2. Angwin, J., Larson, J., Mattu, S., Kirchner, L.: Machine bias risk assessments in criminal sentencing. ProPublica, May 23 (2016)
3. Arnold, M., Bellamy, R.K., Hind, M., Houde, S., Mehta, S., Mojsilović, A., Nair, R., Ramamurthy, K.N., Olteanu, A., Piorkowski, D., et al.: Factsheets: Increasing trust in ai services through supplier's declarations of conformity. IBM J. Res. Dev. **63**(4/5), 6–1 (2019)

4. Bellamy, R.K., et al.: Ai fairness 360: An extensible toolkit for detecting, understanding, and mitigating unwanted algorithmic bias. arXiv preprint arXiv:1810.01943 (2018)
5. Bellù, L.G., Liberati, P.: Inequality analysis: The gini index. FAO, EASYPol Module 40 (2006)
6. Berk, R., Heidari, H., Jabbari, S., Kearns, M., Roth, A.: Fairness in criminal justice risk assessments: The state of the art. Sociological Methods & Research p. 0049124118782533 (2018)
7. Blake, C.: Cj merz uci repository of machine learning databases. University of California at Irvine (1998)
8. Calders, T., Verwer, S.: Three naive bayes approaches for discrimination-free classification. Data Min. Knowl. Disc. $21(2)$, 277–292 (2010)
9. Chouldechova, A.: Fair prediction with disparate impact: A study of bias in recidivism prediction instruments. Big data $5(2)$, 153–163 (2017)
10. Chouldechova, A., Roth, A.: The frontiers of fairness in machine learning. arXiv preprint arXiv:1810.08810 (2018)
11. Corbett-Davies, S., Pierson, E., Feller, A., Goel, S., Huq, A.: Algorithmic decision making and the cost of fairness. In: Proceedings of the 23rd ACM SIGKDD International Conference on Knowledge Discovery and Data Mining, pp. 797–806. ACM (2017)
12. Dua, D., Graff, C.: UCI machine learning repository (2017), http://archive.ics.uci.edu/ml
13. Dwork, C., Hardt, M., Pitassi, T., Reingold, O., Zemel, R.: Fairness through awareness. In: Proceedings of the 3rd Innovations in Theoretical Computer Science Conference, pp. 214–226. ACM (2012)
14. Dwork, C., Ilvento, C.: Fairness under composition. arXiv preprint arXiv:1806.06122 (2018)
15. Dwork, C., Ilvento, C.: Group fairness under composition (2018)
16. Friedler, S.A., Scheidegger, C., Venkatasubramanian, S.: On the (im) possibility of fairness. arXiv preprint arXiv:1609.07236 (2016)
17. Gajane, P., Pechenizkiy, M.: On formalizing fairness in prediction with machine learning. arXiv preprint arXiv:1710.03184 (2017)
18. Hardt, M., Price, E., Srebro, N.: Equality of opportunity in supervised learning. Adv. Neural. Inf. Process. Syst. 29, 3315–3323 (2016)
19. Kamishima, T., Akaho, S., Sakuma, J.: Fairness-aware learning through regularization approach. In: 2011 IEEE 11th International Conference on Data Mining Workshops, pp. 643–650. IEEE (2011)
20. Kearns, M., Roth, A., Sharifi-Malvajerdi, S.: Average individual fairness: algorithms, generalization and experiments. arXiv preprint arXiv:1905.10607 (2019)
21. Kleinberg, J., Mullainathan, S., Raghavan, M.: Inherent trade-offs in the fair determination of risk scores. arXiv preprint arXiv:1609.05807 (2016)
22. Lahoti, P., Gummadi, K.P., Weikum, G.: ifair: learning individually fair data representations for algorithmic decision making. In: 2019 IEEE 35th International Conference on Data Engineering (ICDE), pp. 1334–1345. IEEE (2019)
23. Mitchell, S., Potash, E., Barocas, S.: Prediction-based decisions and fairness: a catalogue of choices, assumptions, and definitions. arXiv preprint arXiv:1811.07867 (2018)
24. Narayanan, A.: Translation tutorial: 21 fairness definitions and their politics. In: Proceedings of Conference on Fairness, Accountability, and Transparency, New York, USA (2018)

25. Ritov, Y., Sun, Y., Zhao, R.: On conditional parity as a notion of non-discrimination in machine learning. arXiv preprint arXiv:1706.08519 (2017)
26. Roscher, R., Bohn, B., Duarte, M.F., Garcke, J.: Explainable machine learning for scientific insights and discoveries. Ieee Access **8**, 42200–42216 (2020)
27. Rothblum, G.N., Yona, G.: Probably approximately metric-fair learning. arXiv preprint arXiv:1803.03242 (2018)
28. Shorrocks, A.F.: Inequality decomposition by population subgroups. Econometrica: Journal of the Econometric Society, pp. 1369–1385 (1984)
29. Speicher, T., et al.: A unified approach to quantifying algorithmic unfairness: Measuring individual & group unfairness via inequality indices. In: Proceedings of the 24th ACM SIGKDD International Conference on Knowledge Discovery & Data Mining, pp. 2239–2248. ACM (2018)
30. Verma, S., Ernst, M., Just, R.: Removing biased data to improve fairness and accuracy. arXiv preprint arXiv:2102.03054 (2021)
31. Verma, S., Rubin, J.: Fairness definitions explained. In: 2018 IEEE/ACM International Workshop on Software Fairness (FairWare), pp. 1–7. IEEE (2018)
32. Wing, J.M.: Trustworthy ai. Commun. ACM **64**(10), 64–71 (2021)
33. Zafar, M.B., Valera, I., Gomez Rodriguez, M., Gummadi, K.P.: Fairness beyond disparate treatment & disparate impact: learning classification without disparate mistreatment. In: Proceedings of the 26th International Conference on World Wide Web, pp. 1171–1180. International World Wide Web Conferences Steering Committee (2017)
34. Zemel, R., Wu, Y., Swersky, K., Pitassi, T., Dwork, C.: Learning fair representations. In: International Conference on Machine Learning, pp. 325–333. PMLR (2013)

Do Minimal Complexity Least Squares Support Vector Machines Work?

Shigeo Abe[✉]

Kobe University, Rokkodai, Nada, Kobe, Japan
abe@kobe-u.ac.jp
http://www2.kobe-u.ac.jp/ abe

Abstract. The minimal complexity support vector machine is a fusion of the support vector machine (SVM) and the minimal complexity machine (MCM), and results in maximizing the minimum margin and minimizing the maximum margin. It works to improve the generalization ability of the L1 SVM (standard SVM) and LP (Linear Programming) SVM. In this paper, we discuss whether it also works for the LS (Least Squares) SVM. The minimal complexity LS SVM (MLS SVM) is trained by minimizing the sum of squared margin errors and minimizing the maximum margin. This results in solving a set of linear equations and a quadratic program, alternatingly. According to the computer experiments for two-class and multiclass problems, the MLS SVM does not outperform the LS SVM for the test data although it does for the cross-validation data.

1 Introduction

A classifier is designed to achieve high generalization ability for unknown data by maximizing class separability. The support vector machine (SVM) [1,2] realizes this by maximizing the minimum margin, where a margin of a data sample is defined as its distance from the separating hyperplane. Although the SVM works relatively well for a wide range of applications, there is still a room for improvement. Therefore, in addition to maximizing the minimum margin, controlling the margin distribution is considered. One approach controls the low order statistics [3–8]. In [5], a large margin distribution machine (LDM) was proposed, in which the average margin is maximized and the margin variance is minimized. Because the LDM includes an additional hyperparameter compared to the SVM, in [6,7], the unconstrained LDM (ULDM) was proposed, which has the same number of hyperparameters as the SVM. The least squares SVM (LS SVM) [3,4] is consider to be based on low order statistics because it minimizes the sum of squared margin errors.

Another approach [9–15] minimizes the VC (Vapnik-Chervonenkis) dimension [1]. In [9], the minimal complexity machine (MCM) that minimizes the VC dimension was proposed, which is reduced to minimizing the sum of margin errors and minimizing the maximum margin. According to the analysis in [10],

N. El Gayar et al. (Eds.): ANNPR 2022, LNAI 13739, pp. 53–64, 2023.
https://doi.org/10.1007/978-3-031-20650-4_5

however, the solution of the MCM was shown to be non-unique and unbounded. These disadvantages can be solved by introducing the regularization term into the MCM, which is a fusion of the LP (Linear Programming) SVM and the MCM called MLP SVM. The soft upper-bound minimal complexity LP SVM (SLP SVM) [14] is a soft upper-bound version of the MLP SVM. The ML1 SVM [11,12] is the fusion of the MCM and the standard SVM (L1 SVM) and the SL1 SVM [15] is a soft upper-bound version of the ML1 SVM. According to the computer experiments, in general, the fusions, i.e., minimization of the maximum margin in the SVMs, improved the generalization ability of the base classifiers, and the ML1$_v$ SVM, which is a variant of the ML1 SVM performed best.

In this paper, we discuss whether the idea of minimizing the VC dimension, i.e., minimizing the maximum margin, also works for the LS SVM, which controls the margin distribution by the second order statistics. We formulate the minimal complexity LS SVM (MLS SVM) by minimizing the maximum margin as well as maximizing the minimum margin in the LS SVM framework. We derive the dual form of the MLS SVM and the training method that trains the MLS SVM alternatingly by matrix inversion and by the SMO (Sequential Minimal Optimization) based Newton's method [16]. By computer experiments, we show whether the MLS SVM performs better than the LS SVM.

In Sect. 2, we discuss the architecture of the MLS SVM and derive its dual problem. Then we discuss the training method of the MLS SVM. And in Sect. 3, we compare generalization performance of the MLS SVM with the LS SVM and other SVM-based classifiers using two-class and multiclass problems.

2 Minimal Complexity Least Squares Support Vector Machines

In this section we discuss the architecture of the MLS SVM, the KKT conditions, and a training method.

2.1 Architecture

For a two-class problem, we consider the following decision function:

$$D(\mathbf{x}) = \mathbf{w}^\top \boldsymbol{\phi}(\mathbf{x}) + b, \tag{1}$$

where $\mathbf{w}$ is the l-dimensional vector, b is the bias term, and $\boldsymbol{\phi}(\mathbf{x})$ is the l-dimensional vector that maps m-dimensional vector $\mathbf{x}$ into the feature space. If $D(\mathbf{x}) > 0$, $\mathbf{x}$ is classified into Class 1 and if $D(\mathbf{x}) < 0$, Class 2.

We introduce the idea of minimizing the VC dimension into the LS SVM: we minimize the maximum margin as well as maximizing the minimum margin.

The minimal complexity LS SVM (MLS SVM) is formulated as follows:

$$\min \quad \frac{1}{2}\mathbf{w}^\top\mathbf{w} + \frac{C}{2}\sum_{i=1}^{M}\xi_i^2 + C_h\left(h^+ + h^-\right) \tag{2}$$

$$\text{s.t.} \quad y_i\left(\mathbf{w}^\top\boldsymbol{\phi}(\mathbf{x}_i) + b\right) = 1 - \xi_i \quad \text{for} \quad i = 1,\ldots,M, \tag{3}$$

$$h_i \geq y_i\left(\mathbf{w}^\top\boldsymbol{\phi}(\mathbf{x}_i) + b\right) \quad \text{for} \quad i = 1,\ldots,M, \tag{4}$$

$$h^+ \geq 1, \quad h^- \geq 1, \tag{5}$$

where $(\mathbf{x}_i, y_i)$ $(i = 1,\ldots,M)$ are M training input-output pairs, $y_i = 1$ if $\mathbf{x}_i$ belong to Class 1, and $y_i = -1$ if Class 2, ξ_i are the slack variables for $\mathbf{x}_i$, C is the margin parameter, h^+ and h^- are the upper bounds for the Classes 1 and 2, respectively, and $h_i = h^+$ for $y_i = 1$ and $h_i = h^-$ for $y_i = -1$. Here, if $\xi_i \geq 1$, $\mathbf{x}_i$ is misclassified and otherwise, $\mathbf{x}_i$ is correctly classified. Unlike L1 or L2 SVMs, ξ_i can be negative. The first term in the objective function is the reciprocal of the squared margin divided by 2, the second term is to control the number of misclassifications, and C controls the tradeoff between the first and second terms. The third term works to minimize the maximum margin. Parameter C_h controls the upper bounds h^+ and h^-.

If we delete (4), (5), and the third term in (2), we obtain the LS SVM. And if in the LS SVM we replace the equality constraints in (3) into the inequality constraints ($\geq$) and the square sum of slack variables in (2) into the linear sum multiplied by 2, we obtain the L1 SVM, which is a standard SVM.

In the following, we derive the dual problem of the above optimization problem.

Introducing the Lagrange multipliers α_i, $\alpha_{M+i}\,(\geq 0)$, $\eta^+\,(\geq 0)$, and $\eta^-\,(\geq 0)$ into (2) to (5), we obtain the unconstrained objective function:

$$Q(\mathbf{w}, b, \boldsymbol{\alpha}, \boldsymbol{\xi}, h^+, h^-, \eta^+, \eta^-)$$

$$= \frac{1}{2}\mathbf{w}^\top\mathbf{w} + \frac{C}{2}\sum_{i=1}^{M}\xi_i^2 + C_h\left(h^+ + h^-\right) - \sum_{i=1}^{M}\alpha_i\left(y_i(\mathbf{w}^\top\boldsymbol{\phi}(\mathbf{x}_i) + b) - 1 + \xi_i\right),$$

$$- \sum_{i=1}^{M}\alpha_{M+i}\left(h_i - y_i(\mathbf{w}^\top\boldsymbol{\phi}(\mathbf{x}_i) + b)\right) - \eta^+\left(h^+ - 1\right) - \eta^-\left(h^- - 1\right) \tag{6}$$

where $\boldsymbol{\alpha} = (\alpha_1,\ldots,\alpha_M,\alpha_{M+1},\ldots,\alpha_{2M})^\top$, and $\boldsymbol{\xi} = (\xi_1,\ldots,\xi_M)^\top$.

Taking the partial derivatives of (6) with respect to $\mathbf{w}$, b, $\boldsymbol{\xi}$, h^+, and h^- and equating them to zero, together with the equality constraints (3), we obtain the optimal conditions as follows:

$$\mathbf{w} = \sum_{i=1}^{M} y_i \left(\alpha_i - \alpha_{M+i}\right) \phi(\mathbf{x}_i), \tag{7}$$

$$\sum_{i=1}^{M} y_i \left(\alpha_i - \alpha_{M+i}\right) = 0, \tag{8}$$

$$\alpha_i = C\xi_i \quad \text{for} \quad i = 1, \ldots, M, \tag{9}$$

$$C_h = \sum_{i=1, y_i=1}^{M} \alpha_{M+i} + \eta^+, \quad C_h = \sum_{i=1, y_i=-1}^{M} \alpha_{M+i} + \eta^-, \tag{10}$$

$$y_i \left(\mathbf{w}^\top \phi(\mathbf{x}_i) + b\right) - 1 + \xi_i = 0 \quad \text{for} \quad i = 1, \ldots, M, \tag{11}$$

$$\alpha_{M+i} \left(h_i - y_i \left(\mathbf{w}^\top \phi(\mathbf{x}_i) + b\right)\right) = 0, \quad \alpha_{M+i} \ge 0 \quad \text{for} \quad i = 1, \ldots, M, \tag{12}$$

$$\eta^+ \left(h^+ - 1\right) = 0, \quad \eta^+ \ge 0, \quad \eta^- \left(h^- - 1\right) = 0, \quad \eta^- \ge 0. \tag{13}$$

From (9), unlike L1 or L2 SVMs, α_i can be negative.

Now, we derive the dual problem. Substituting (7) and (8) into (6), we obtain the objective function with respect to $\boldsymbol{\alpha}$, η^+, and η^-. Thus, we obtain the following dual problem:

$$\max \quad Q(\boldsymbol{\alpha}, \eta^+, \eta^-) = \sum_{i=1}^{M} \alpha_i - \frac{1}{2} \sum_{i,j=1}^{M} y_i \left(\alpha_i - \alpha_{M+i}\right)$$

$$\times y_j \left(\alpha_j - \alpha_{M+j}\right) K(\mathbf{x}_i, \mathbf{x}_j) - \frac{1}{2C} \sum_i^{M} \alpha_i^2 + \eta^+ + \eta^-, \tag{14}$$

$$\text{s.t.} \quad \sum_{i=1}^{M} y_i \left(\alpha_i - \alpha_{M+i}\right) = 0, \tag{15}$$

$$C_h \ge \sum_{i=1, y_i=1}^{M} \alpha_{M+i}, \quad C_h \ge \sum_{i=1, y_i=-1}^{M} \alpha_{M+i}, \tag{16}$$

$$\alpha_{M+i} \ge 0 \quad \text{for} \quad i = 1, \ldots, M, \tag{17}$$

where $K(\mathbf{x}, \mathbf{x}')$ is the kernel and $K(\mathbf{x}, \mathbf{x}') = \phi^\top(\mathbf{x}) \phi(\mathbf{x}')$. Similar to the SVM, defining $K(\mathbf{x}, \mathbf{x}')$, we can avoid the explicit treatment of variables in the feature space.

In the above optimization problem, if we delete $(\alpha_{M+1}, \ldots, \alpha_{2M})$, η^+, η^-, and their related terms, we obtain the LS SVM.

Similar to the ML1$_v$ SVM [12], we assume that $\eta^+ = \eta^- = 0$. This means that $h^+ \geq 1$ and $h^- \geq 1$. Then the optimization problem reduces to

$$\max \quad Q(\boldsymbol{\alpha}) = \sum_{i=1}^{M} \alpha_i - \frac{1}{2} \sum_{i,j=1}^{M} (\alpha_i - \alpha_{M+i})(\alpha_j - \alpha_{M+j}) y_i y_j K(\mathbf{x}_i, \mathbf{x}_j)$$

$$- \frac{1}{2C} \sum_{i=1}^{M} \alpha_i^2, \tag{18}$$

$$\text{s.t.} \quad \sum_{i=1}^{M} y_i \alpha_i = 0, \tag{19}$$

$$C_h = \sum_{i=1,y_i=1}^{M} \alpha_{M+i} = \sum_{i=1,y_i=-1}^{M} \alpha_{M+i}, \tag{20}$$

$$\alpha_{M+i} \geq 0 \quad \text{for} \quad i = 1, \ldots, M. \tag{21}$$

Notice that because of (20), (15) reduces to (19).

We decompose the above optimization problem into two subprograms:

1. Subproblem 1 Solving the problem for $\alpha_1, \ldots, \alpha_M$ and b fixing $\alpha_{M+1}, \ldots, \alpha_{2M}$:

$$\max \quad Q(\boldsymbol{\alpha}^0) = \sum_{i=1}^{M} \alpha_i - \frac{1}{2} \sum_{i,j=1}^{M} (\alpha_i - \alpha_{M+i})(\alpha_j - \alpha_{M+j}) y_i y_j K(\mathbf{x}_i, \mathbf{x}_j)$$

$$- \frac{1}{2C} \sum_{i=1}^{M} \alpha_i^2, \tag{22}$$

$$\text{s.t.} \quad \sum_{i=1}^{M} y_i \alpha_i = 0, \tag{23}$$

where $\boldsymbol{\alpha}^0 = (\alpha_1, \ldots, \alpha_M)^\top$.

2. Subproblem 2 Solving the problem for $\alpha_{M+1}, \ldots, \alpha_{2M}$ fixing $\boldsymbol{\alpha}^0$ and b:

$$\max \quad Q(\boldsymbol{\alpha}^M) = -\frac{1}{2} \sum_{i,j=1}^{M} (\alpha_i - \alpha_{M+i})(\alpha_j - \alpha_{M+j}) y_i y_j K(\mathbf{x}_i, \mathbf{x}_j) \tag{24}$$

$$\text{s.t.} \quad C_h = \sum_{i=1,y_i=1}^{M} \alpha_{M+i} = \sum_{i=1,y_i=-1}^{M} \alpha_{M+i}, \tag{25}$$

$$\alpha_{M+i} \geq 0 \quad \text{for} \quad i = 1, \ldots, M, \tag{26}$$

where $\boldsymbol{\alpha}^M = (\alpha_{M+1}, \ldots, \alpha_{2M})^\top$.

We must notice that as the value of C_h approaches zero, the MLS SVM reduces to the LS SVM. Therefore, for a sufficiently small value of C_h, the MLS SVM and LS SVM behave similarly.

We consider solving the above subproblems alternatingly.

Here, because of (25), if we modify α_{M+i}, another α_{M+j} belonging to the same class must be modified. Therefore, $\boldsymbol{\alpha}^M$ can be updated per class.

2.2   Solving Subproblem 1

Variables $(\alpha_1, \ldots, \alpha_M)$ and b can be solved for using (7), (9), (11), and (23) by matrix inversion. Substituting (7) and (9) into (11) and expressing it and (23) in matrix form, we obtain

$$\begin{pmatrix} \Omega & 1 \\ 1^\top & 0 \end{pmatrix} \begin{pmatrix} \boldsymbol{\alpha}' \\ b \end{pmatrix} = \begin{pmatrix} \mathbf{d}_1 \\ 0 \end{pmatrix}, \tag{27}$$

or

$$\Omega \boldsymbol{\alpha}' + \mathbf{1} b = \mathbf{d}_1, \tag{28}$$
$$\mathbf{1}^\top \boldsymbol{\alpha}' = 0, \tag{29}$$

where $\mathbf{1}$ is the M-dimensional vector and

$$\boldsymbol{\alpha}' = (y_1\, \alpha_1, \ldots, y_M\, \alpha_M)^\top \tag{30}$$
$$\Omega_{ij} = K(\mathbf{x}_i, \mathbf{x}_j) + \frac{\delta_{ij}}{C}, \tag{31}$$
$$\mathbf{d}_1 = (d_{11}, \ldots, d_{1M})^\top, \tag{32}$$
$$d_{1i} = y_i + \sum_{j=1}^{M} y_j\, \alpha_{M+j}\, K(\mathbf{x}_i, \mathbf{x}_j), \tag{33}$$
$$\mathbf{1} = (1, \ldots, 1)^\top, \tag{34}$$

where $\delta_{ij} = 1$ for $i = j$, and $\delta_{ij} = 0$ for $i \neq j$.

If $\boldsymbol{\alpha}^M = \mathbf{0}$, (27) reduces to solving the LS SVM.

Subproblem 1 is solved by solving (27) for $\boldsymbol{\alpha}^0$ and b as follows. Because of $1/C\,(> 0)$ in the diagonal elements of Ω, Ω is positive definite. Therefore,

$$\boldsymbol{\alpha}' = \Omega^{-1}(\mathbf{d}_1 - \mathbf{1}\,b). \tag{35}$$

Substituting (35) into (29), we obtain

$$b = (\mathbf{1}^\top \Omega^{-1} \mathbf{1})^{-1} \mathbf{1}^\top \Omega^{-1} \mathbf{d}_1. \tag{36}$$

Thus, substituting (36) into (35), we obtain $\boldsymbol{\alpha}'$.

2.3   Solving Subproblem 2

Subproblem 2 needs to be solved iteratively. We derive the KKT (Karush-Kuhn-Tucker) conditions for Subproblem 2 for the convergence check. Because of the space limitation, we skip the detailed training method based on the SMO (Sequential Minimal Optimization) combined with Newton's method [16].

For Subprogram 2, training is converged if the KKT optimality condition (12) is satisfied. Substituting (7) and (9) into (12), we obtain the following KKT conditions:

$$\alpha_{M+i}\,(h_i + y_i\,F_i - y_i\,b) = 0 \quad \text{for} \quad i = 1,\dots,M, \tag{37}$$

where

$$F_i = -\sum_{j=1}^{M} y_j(\alpha_j - \alpha_{M+j})K(\mathbf{x}_i,\mathbf{x}_j). \tag{38}$$

Here the value of b is determined in Subprogram 1.

KKT Conditions. We can classify the conditions of (37) into the following two cases:

1. $\alpha_{M+i} = 0$. From $h_i \geq -y_i\,F_i + y_i\,b$,

$$F_i \geq b - h^+ \text{ for } y_i = 1, \ \ b + h^- \geq F_i \text{ for } y_i = -1. \tag{39}$$

2. $C_h \geq \alpha_{M+i} > 0$. From $h_i = -y_i\,F_i + y_i\,b$,

$$b - h^+ = F_i \text{ for } y_i = 1, \ \ b + h^- = F_i \text{ for } y_i = -1. \tag{40}$$

Then the KKT conditions for (37) are simplified as follows:

$$\bar{F}_i^{\,+} \geq b - h^+ \geq \tilde{F}_i^{\,+} \text{ for } y_i = 1,$$
$$\bar{F}_i^{\,-} \geq b + h^- \geq \tilde{F}_i^{\,-} \text{ for } y_i = -1, \quad \text{for} \quad i = 1,\dots,M, \tag{41}$$

where

$$\bar{F}_i^{\,+} = F_i \ \text{ if } \ \alpha_{M+i} \geq 0, \quad \tilde{F}_i^{\,+} = F_i \ \text{ if } \ \alpha_{M+i} > 0, \tag{42}$$
$$\bar{F}_i^{\,-} = F_i \ \text{ if } \ \alpha_{M+i} > 0, \quad \tilde{F}_i^{\,-} = F_i \ \text{ if } \ \alpha_{M+i} \geq 0. \tag{43}$$

To detect the violating variables, we define b_{up}^s and b_{low}^s as follows:

$$b_{\text{up}}^s = \min_i \bar{F}_i^{\,s}, \quad b_{\text{low}}^s = \max_i \tilde{F}_i^{\,s}, \tag{44}$$

where $s = +, -$, $b^+ = b - h^+$, and $b^- = b + h^-$.

If the KKT conditions are satisfied,

$$b_{\text{up}}^s \geq b_{\text{low}}^s. \tag{45}$$

To speed up training we consider that training is converged if

$$\max_{s=+,-} b_{\text{low}}^s - b_{\text{up}}^s \leq \tau, \tag{46}$$

where τ is a small positive parameter.

And the upper bounds are estimated to be

$$h_{\text{e}}^+ = b - \frac{1}{2}(b_{\text{up}}^+ + b_{\text{low}}^+), \ \ h_{\text{e}}^- = -b + \frac{1}{2}(b_{\text{up}}^- + b_{\text{low}}^-). \tag{47}$$

2.4   Training Procedure

In the following we show the training procedure of the MLS SVM.

1. (Solving Subprogram 1) Solve (27) for $\boldsymbol{\alpha}^0$ and b fixing $\boldsymbol{\alpha}^M$ with the solution obtained in Step 2. Initially we set $\boldsymbol{\alpha}^M = \mathbf{0}$.
2. (Solving Subprogram 2) Solve (24)–(26) for $\boldsymbol{\alpha}^M$ fixing $\boldsymbol{\alpha}^0$ and b with the solution obtained in Step 1. Initially, we set one α_{M+i} in each class to C_h.
3. (Convergence check) If (46) is satisfied, finish training. Otherwise go to Step 1.

The objective function $Q(\boldsymbol{\alpha})$ is monotonic during training: In Step 1, the objective function is maximized with the fixed $\boldsymbol{\alpha}^M$. Therefore the objective function is non-decreasing after $\boldsymbol{\alpha}^0$ and b are corrected. In Step 2, the objective function is maximized with the fixed $\boldsymbol{\alpha}^0$ and b. Therefore, the objective function is also non-decreasing after $\boldsymbol{\alpha}^M$ is corrected. In Step 2, so long as (45) is not satisfied, the objective function is increased by correcting $\boldsymbol{\alpha}^M$. Therefore, the training stops within finite steps.

The hyperparameter values of γ, C, and C_h are determined by cross-validation. To make the accuracy improvement over the LS SVM clear, in the following performance evaluation, we determined the values of γ and C, with $C_h = 0$, i.e., using the LS SVM. After they were determined, we determined the C_h value of MLS SVM. By this method, we can make the accuracy of the MLS SVM at least by cross-validation not lower than that of the LS SVM, if the smallest value of C_h in cross-validation is sufficiently small.

3   Performance Evaluation

We evaluated whether the idea of minimizing the VC-dimension, i.e., minimizing the maximum margin works to improve the generalization ability of the LS SVM. As classifiers we used the MLS SVM, LS SVM, ML1$_v$ SVM, which is a variant of ML1 SVM, L1 SVM, and ULDM. As a variant of the MLS SVM, we used an early stopping MLS SVM, MLS$_e$ SVM, which terminates training when the Subprogram 2 converges after matrix inversion for Subprogram 1 is carried out. This was to check whether early stopping improves the generalization ability when overfitting occurs.

To make comparison fair we determined the values of the hyperparameters by fivefold cross-validation of the training data, trained the classifiers with the selected hyperparameter values, and tested the accuracies for the test data. (Because of the computational cost we did not use nested (double) cross-validation.) We used two-class and multiclass problems used in [15]. The two-class problems have 100 or 20 pairs of training and test data sets and the multiclass problems, one, each. In cross-validation, the candidate values for γ and C were the same as those discussed in [15]. Those for C_h in the MLS SVM and MLS$_e$ SVM were $\{0.001, 0.01, 0.1, 1, 10, 50, 100, 500\}$ instead of $\{0.1, 1, 10, 50, 100, 500, 1000, 2000\}$ in the ML1$_v$ SVM. This was to avoid deteriorating the cross-validation accuracy in determining the C_h value. In addition, a tie was broken by selecting a smallest value except for MLS_e SVM; for the MLS_e SVM, a largest value was selected so that minimizing the maximum margin worked.

Table 1 shows the average accuracies for the 13 two-class problems. In the first column, in I/Tr/Te, I shows the number of inputs, Tr, the number of training data, and Te, the number of test data. For each problem, the best average accuracy is in bold and the worst, underlined. The average accuracy is accompanied by the standard deviation. The plus sign attached to the accuracy shows that the MLS SVM is statistically better than the associated classifier. Likewise, the minus sign, worse than the associated classifier. The "Average" row shows the average accuracy of the associated classifier for the 13 problems and B/S/W denotes that the associated classifier shows the best accuracy B times, the second best, S times, and the worst accuracy, W times. The "W/T/L" denotes that the MLS SVM is statistically better than the associated classifier W times, comparable to, T times, and worse than, L times.

From the Average measure, the ULDM performed best and the MLS_e SVM, the worst. And both the MLS SVM and MLS_e SVM were inferior to the LS SVM. From the B/S/W measure, also the ULDM was the best and the LS SVM was the second best. From the W/T/L measure, the MLS SVM was better than the MLS_e SVM and comparable or almost comparable to the LS SVM, $ML1_v$ SVM, and ULDM. Although the MLS SVM was statistically comparable to or better than other classifiers, from the Average measure, it was inferior to the LS SVM. To investigate, why this happened, we compared the average accuracy obtained by cross-validation, which is shown in Table 2. From the table, the MLS SVM showed the best average accuracies for all the problems. This shows that the idea of minimizing the maximum margin worked for the MLS SVM at least for the cross-validation accuracies. But from Table 1, the MLS SVM was better than or equal to the LS SVM for only three problems: the diabetes, flare-solar, and splice problems. This shows that in most cases overfitting occurred for the MLS SVM. On the other hand, the MLS_e SVM was inferior to the LS SVM except for the test data accuracy of the titanic problem. Thus, in most cases, inferior performance was caused by underfitting.

Table 4 shows the accuracies of the test data for the multiclass problems. The original MNIST data set has 6000 data points per class and it is difficult to train the low order statistic-based classifiers by matrix inversion. Therefore, to reduce the cross-validation time, we switched the roles of training and test data sets for the MNIST problem and denote it as MNIST (r). From the Average measure, the $ML1_v$ SVM performed best, the ULDM the second best, and the MLS SVM and MLS_e SVM, worst. From the B/S/W measure, the MLS_e SVM was the best and the MLS SVM the worst. For the MLS_e SVM, the accuracy for the thyroid problem was the worst. Comparing the MLS_e SVM and LS SVM, the MLS_e SVM performed better than the LS SVM six times, but the MLS SVM, only once. Therefore, the MLS_e SVM performed better than the LS SVM but MLS SVM did not.

Table 1. Average accuracies of the test data for the two-class problems

Problem I/Tr/Te	MLS	MLS$_e$	LS	ML1$_v$	L1	ULDM
Banana 2/400/4900	89.16 ± 0.68	89.02 ± 0.79	**89.17** ± 0.66	89.13 ± 0.70	**89.17** ± 0.74	89.12 ± 0.69
Cancer 9/200/77	72.99 ± 4.66	71.01$^+$ ± 4.38	73.13 ± 4.68	73.14 ± 4.38	72.99 ± 4.49	**73.70** ± 4.42
Diabetes 8/468/300	76.21 ± 2.01	74.76$^+$ ± 2.77	76.19 ± 2.00	76.36 ± 1.84	76.23 ± 1.80	**76.51** ± 1.95
Flare-solar 9/666/400	66.25 ± 1.98	63.62$^+$ ± 2.65	66.25 ± 1.98	**66.99**$^-$ ± 2.16	**66.99**$^-$ ± 2.12	66.28 ± 2.05
German 20/700/300	76.00 ± 2.28	74.72$^+$ ± 3.31	76.10 ± 2.10	75.88 ± 2.18	76.01 ± 2.12	**76.12** ± 2.30
Heart 13/170/100	82.43 ± 3.53	82.35 ± 3.61	82.49 ± 3.60	**82.89** ± 3.33	82.72 ± 3.40	82.57 ± 3.64
Image 18/1300/1010	97.50 ± 0.57	97.14$^+$ ± 0.52	**97.52** ± 0.54	97.28 ± 0.46	97.16$^+$ ± 0.41	97.16 ± 0.68
Ringnorm 20/400/7000	98.18 ± 0.35	97.29$^+$ ± 1.56	**98.19** ± 0.33	98.01 ± 1.11	98.14 ± 0.34	98.16 ± 0.35
Splice 60/1000/2175	89.00 ± 0.71	88.93 ± 0.82	88.98 ± 0.70	88.99 ± 0.83	88.89 ± 0.84	**89.16** ± 0.53
Thyroid 5/140/75	95.04 ± 2.56	94.84 ± 2.60	95.08 ± 2.55	95.35 ± 2.48	**95.39** ± 2.43	95.15 ± 2.27
Titanic 3/150/2051	77.30 ± 1.27	77.42 ± 0.78	77.39 ± 0.83	77.42 ± 0.74	77.35 ± 0.80	**77.46** ± 0.91
Twonorm 20/400/7000	97.40 ± 0.28	97.05$^+$ ± 0.60	**97.43** ± 0.27	97.37 ± 0.28	97.38 ± 0.27	97.41 ± 0.26
Waveform 21/400/4600	90.01 ± 0.58	89.32$^+$ ± 1.15	90.05 ± 0.59	89.66$^+$ ± 0.76	89.72$^+$ ± 0.70	**90.18**$^-$ ± 0.54
Average (B/S/W)	85.19 (0/3/1)	84.42 (0/1/12)	85.23 (4/2/0)	85.27 (2/4/0)	85.24 (3/1/0)	**85.31** (6/1/0)
W/T/L	—	8/5/0	0/13/0	1/11/1	2/10/1	0/12/1

Table 2. Average accuracies by cross-validation for the two-class problems

Problem	MLS	MLS$_e$	LS
Banana	**90.60**	90.21	90.50
Cancer	**76.03**	72.77	75.99
Diabetes	**78.19**	76.07	78.15
Flare-solar	**67.38**	63.59	67.36
German	**76.59**	74.36	76.58
Heart	**84.70**	83.99	84.59
Image	**97.39**	97.24	97.38
Ringnorm	**98.65**	97.53	98.60
Splice	**89.02**	89.01	88.94
Thyroid	**97.44**	97.04	97.37
Titanic	**79.49**	78.81	79.45
Twonorm	**98.06**	97.55	97.98
Waveform	**91.06**	90.05	91.00
Average	**86.51**	85.25	86.45
B/W	13/0	0/12	0/1

Table 3. Average accuracies by cross-validation for the multiclass problems

Problem	MLS	MLS$_e$	LS
Numeral	**99.63**	99.51	**99.63**
Thyroid	**95.97**	95.02	**95.97**
Blood cell	**94.83**	94.54	**94.83**
Hiragana-50	99.67	**99.70**	99.67
Hiragana-13	**99.86**	99.83	**99.86**
Hiragana-105	**99.98**	**99.98**	**99.98**
Satimage	92.72	**92.76**	92.72
USPS	**98.46**	98.44	98.44
MNIST(r)	**97.59**	**97.59**	**97.59**
Letter	**97.81**	97.74	**97.81**
Average	**97.65**	97.51	**97.65**
B/W	8/2	4/6	7/3

Now examine the result from the cross-validation accuracies shown in Table 3. The accuracies of the MLS SVM were the same as those of the LS SVM except for the USPS problem. Therefore, from Table 4, the idea of minimizing the maximum margin did not contribute in improving the accuracies of the test data except for the blood cell problem. For the MLS$_e$ SVM, the adverse effect of early stopping occurred for the thyroid problem: the worst accuracy of the test data in Table 4 was caused by underfitting as seen from Table 3. For the remaining problems the adverse effect was small or none.

Table 4. Accuracies of the test data for the multiclass problems

Problem I/C/Tr/Te	MLS	MLS$_e$	LS	ML1$_v$	L1	ULDM
Numeral 12/10/810/820 [2]	99.15	99.39	99.15	**99.76**	**99.76**	99.39
Thyroid 21/3/3772/3428 [17]	95.39	94.57	95.39	97.23	**97.26**	95.57
Blood cell 13/12/3097/3100 [2]	94.29	94.29	94.23	93.65	93.19	**94.61**
Hiragana-50 50/39/4610/4610 [2]	99.20	99.28	**99.48**	99.11	98.98	98.92
Hiragana-13 13/38/8375/8356 [2]	99.87	99.88	99.87	**99.92**	99.76	99.90
Hiragana-105 105/38/8375/8356 [2]	**100.00**	**100.00**	**100.00**	**100.00**	**100.00**	**100.00**
Satimage 36/6/4435/2000 [17]	91.95	**92.30**	91.95	91.85	91.90	92.25
USPS 256/10/7291/2007 [18]	95.42	**95.52**	95.47	95.37	95.27	95.42
MNIST(r) 784/10/10000/60000 [19]	96.98	**97.03**	96.99	96.95	96.55	**97.03**
Letter 16/26/16000/4000 [17]	97.87	97.85	97.88	**98.03**	97.70	97.75
Average	97.01	97.01	97.04	**97.18**	97.04	97.08
B/S/W	1/1/1	4/2/1	2/2/1	4/1/1	3/0/5	3/2/1

For the experiment of the multiclass problems, we compared the accuracies of the classifiers because we had only one training data set and one test data set. It was possible to generate multiple training and test data sets from the original data. However, we avoided this because of long cross-validation time. To strengthen performance comparison, in the future, we would like to compare classifiers statistically using multiple training and test data sets.

4   Conclusions

In this paper we proposed the minimal complexity least squares support vector machine (MLS SVM), which is a fusion of the LS SVM and the minimal complexity machine (MCM). Unlike the ML1$_v$ SVM, which is a fusion of the L1 SVM and the MCM, the MLS SVM did not show an improvement in the accuracy for the test data over the LS SVM although the MLS SVM showed an improvement for the cross-validation accuracy. However, early stopping of the MLS SVM training sometimes showed improvement over the LS SVM.

In the future, we would like to compare performance of classifiers statistically using multiple training and test data sets.

Acknowledgment. This work was supported by JSPS KAKENHI Grant Numbers 19K04441 and 22K04154.

References

1. Vapnik, V.N.: Statistical Learning Theory. Wiley, New York (1998)
2. Abe, S.: Support Vector Machines for Pattern Classification, 2nd edn. Springer, London (2010)
3. Suykens, J.A.K.: Least squares support vector machines for classification and nonlinear modelling. Neural Network World **10**(1–2), 29–47 (2000)
4. Suykens, J.A.K., Van Gestel, T., De Brabanter, J., De Moor, B., Vandewalle, J.: Least Squares Support Vector Machines. World Scientific Publishing, Singapore (2002)
5. Zhang, T., Zhou, Z.-H.: Large margin distribution machine. In: Twentieth ACM SIGKDD Conference on Knowledge Discovery and Data Mining, pp. 313–322 (2014)
6. Abe, S.: Unconstrained large margin distribution machines. Pattern Recogn. Lett. **98**, 96–102 (2017)
7. Abe, S.: Effect of equality constraints to unconstrained large margin distribution machines. In: Pancioni, L., Schwenker, F., Trentin, E. (eds.) ANNPR 2018. LNCS (LNAI), vol. 11081, pp. 41–53. Springer, Cham (2018). https://doi.org/10.1007/978-3-319-99978-4_3
8. Zhang, T., Zhou, Z.: Optimal margin distribution machine. IEEE Trans. Knowl. Data Eng. **32**(6), 1143–1156 (2020)
9. Jayadeva: Learning a hyperplane classifier by minimizing an exact bound on the VC dimension. Neurocomputing **149**, 683–689 (2015)
10. Abe, S.: Analyzing minimal complexity machines. In: Proceedings of International Joint Conference on Neural Networks, pp. 1–8. Budapest, Hungary (2019)
11. Abe, S.: Minimal complexity support vector machines. In: Schilling, F.-P., Stadelmann, T. (eds.) ANNPR 2020. LNCS (LNAI), vol. 12294, pp. 89–101. Springer, Cham (2020). https://doi.org/10.1007/978-3-030-58309-5_7
12. Abe, S.: Minimal complexity support vector machines for pattern classification. Computers **9**, 88 (2020)
13. Jayadeva, Soman, S., Pant, H., Sharma, M.: QMCM: Minimizing Vapnik's bound on the VC dimension. Neurocomputing **399**, 352–360 (2020)
14. Abe, S.: Soft upper-bound minimal complexity LP SVMs. In: Proceedings of International Joint Conference on Neural Networks, pp. 1–7 (2021)
15. Abe, S.: Soft upper-bound support vector machines. In: Proceedings of International Joint Conference on Neural Networks, pp. 1–8 (2022)
16. Abe, S.: Fusing sequential minimal optimization and Newton's method for support vector training. Int. J. Mach. Learn. Cybern. **7**(3), 345–364 (2016)
17. Asuncion, A., Newman, D.J.: UCI machine learning repository (2007). http://www.ics.uci.edu/~mlearn/MLRepository.html
18. USPS Dataset. https://www.kaggle.com/bistaumanga/usps-dataset
19. LeCun, Y., Cortes, C.: The MNIST database of handwritten digits. http://yann.lecun.com/exdb/mnist/

A Review of Capsule Networks
in Medical Image Analysis

Heba El-Shimy[1]([✉])(iD), Hind Zantout[1](iD), Michael Lones[2](iD),
and Neamat El Gayar[1](iD)

[1] Heriot-Watt University, Dubai, UAE
`he12@hw.ac.uk`
[2] Heriot-Watt University, Edinburgh, Scotland

Abstract. Computer-aided diagnosis technologies are gaining increased
focus within the medical field due to their role in assisting physicians in
their diagnostic decision-making through the ability to recognise patterns
in medical images. Such technologies started showing promising results
in their ability to match or outperform physicians in certain specialities
and improve the quality of medical diagnosis. Convolutional neural networks are one state-of-the-art technique to use for disease detection and
diagnosis in medical images. However, capsule networks aim to improve
over these by preserving part-whole relationships between an object and
its sub-components leading to better interpretability, an important characteristic for applications in the medical domain. In this paper, we review
the latest applications of capsule networks in computer-aided diagnosis
from medical images and compare their results with those of convolutional neural networks employed for the same tasks. Our findings support the use of Capsule Networks over Convolutional Neural Networks
for Computer-Aided Diagnosis due to their superiority in performance
but more importantly for their better interpretability and their ability
to achieve such performance on small datasets.

Keywords: Capsule networks · Deep learning · CADx · Computer-
aided detection · Computer-aided diagnosis · Medical imaging

1 Introduction

The concepts of Computer-Aided Detection (CADe) and Computer-Aided Diagnosis (CADx) started developing since the advent of digital image processing
techniques and entered the stage of large-scale, systematic research in the 1980s.
These systems are able to learn patterns in images that could at times be difficult to detect by the human eye and thus can help clinicians detect and diagnose
diseases in medical images. CADe and CADx systems act as a way for physicians to get a second-opinion before making a final decision [1]. Early CADe/x
systems used traditional machine learning (ML) techniques which required a
manual or separate feature extraction step to be performed by a domain-
expert before feeding those features into the model as input. These hand-crafted

N. El Gayar et al. (Eds.): ANNPR 2022, LNAI 13739, pp. 65–80, 2023.
https://doi.org/10.1007/978-3-031-20650-4_6

features resulted in a limitation for feature-based ML techniques due to possible human perceptual biases [2]. In recent years, there has been a growing interest in Deep Learning (DL) techniques and the possibilities that they carry for utilising CADe/x systems in the field of medical image analysis. One of the significant advantages of DL techniques over traditional ML is that they do not require prior feature extraction as that is done inherently during model training. DL models can analyse images in more detail and can train millions of parameters which often makes them superior to traditional ML techniques, at least in terms of predictive accuracy, when a lot of data is available.

Tasks that are common in CADe/x systems are image classification, image segmentation or object localisation within a frame or image. All of these tasks have been successfully accomplished using Convolutional Neural Networks (CNNs) with remarkable performance in terms of high accuracy and low false negatives. CNNs can handle high dimensional data such as images, and they are capable of efficiently extracting and learning features that are determinant for the relevant classes. Countless architectures have been proposed for CNNs, but the basic building blocks are convolutional and pooling layers. While convolutional layers act as the feature extractors and provide translational equivariance, the pooling layers act as dimensionality reduction steps via feature aggregation, thus providing spatial invariance. It is this pooling process that Sabour et al. [3] argued causes loss of valuable information and should be replaced by another mechanism leading the authors to propose Capsule Networks (CapsNets) as a novel approach to overcome the limitations of CNNs.

In this paper, we review the most recent literature about applications of CapsNets in CADe/x systems for the detection and diagnosis of diseases in medical images. Our main contributions are: 1) exploring the latest publications on CapsNets and selecting for review the ones that match the criteria of being applied in a medical domain for diagnostic purposes and of using image data, i.e., techniques using signals, text and other formats as input data were excluded; 2) we compare CapsNets performance with that of CNNs where possible; and 3) we draw a conclusion of the best technique to use for CADe/x systems, comment on the limitations of the chosen technique and recommend directions for future research. To the best of our knowledge, this work is the only review of CapsNets applications in the medical domain. The rest of this paper is structured as follows: Sect. 2 will provide a theoretical background on CapsNets and their advantages over CNNs. In Sect. 3, we report and summarise the various approaches in the literature utilising CapsNets for CADe/x tasks organised by type of disease and affected organ. In Sect. 4, we discuss our findings and comment on the potential and limitations of CapsNets based on our research, and finally, Sect. 5 concludes the paper.

2   Background on Capsule Networks

2.1   Limitations of CNNs

It has been shown in numerous studies that CNNs have important limitations. For instance, they generally require large datasets in order to achieve acceptable performance; this is particularly significant within a medical context, where the collection of large datasets is often clinically infeasible. Another important limitation is their inability to understand global objects, shapes and poses; hence, they tend to perform less well with shape-based tasks than with texture-based tasks [4]. Additionally, [3] argue that there is some loss of information that happens in CNN pooling layers, as they ignore all but areas with the highest response to the convolution operation in layers that preceded the pooling layer. Pooling is an efficient dimensionality reduction method, but it comes at the cost of some information loss, as noted by [3] and [5]. Finally, CNNs are not intrinsically interpretable and are commonly referred to as "black boxes" [6,7]. Moreover, techniques used to explain CNNs are not always accurate or faithful to the original model [6]. This *opaqueness* of CNNs makes them unfavourable for medical applications where the clinicians need to rationalise a model's decision and provide the patient's right to an explanation for any medical decisions. In conclusion, CNNs can understand objects at the local level, but not as a whole or how the detected features spatially relate to each other.

2.2   Advantages of Capsule Networks

In computer graphics, a program starts with the instantiation parameters of an object, such as the position, size, orientation, deformation, velocity, hue, texture, etc., and uses these parameters to draw the object. CapsNets are able to invert this process by learning from objects in input images their instantiation parameters. CapsNets can thus achieve equivariance. Another advantage of CapsNets is the dynamic routing algorithm that acts as a disentanglement technique to "explain-away" part-whole relationships using a structure that resembles a *parse-tree*. Dynamic routing in capsule networks can be similar to the purpose of self-attention in transformers in trying to understand part-whole relationships [8]. Attention maps in transformers can be mapped to routing coefficients in CapsNets; the difference is that in transformers the attention is computed top-down while in CapsNets the routing coefficients are computed bottom-up [9]. This quality of CapsNet of understanding part-whole relationships allows for better interpretability that is intrinsic in the network and which allows rationalising a CapsNet model's decision, a process that is crucial to medical applications.

2.3   Capsule Network Architecture

A CapsNet consists of two components: an encoder and a decoder, as shown in Fig. 1. The first layer in the network is a regular convolutional layer, as in any CNN, followed by a second convolution operation where the output feature maps are reshaped into blocks of width, height and depth, and containing groups of neurons called *capsules* which are the basic building block for CapsNets. The intuition behind having these capsules is to learn the instantiation parameters of entities in an image. The output of each capsule is a vector containing the encoded properties learnt about a single entity. The probability of the existence of an entity is represented by the length of the output vector. A *squashing* function (Eq. 1) is used on each capsule to introduce non-linearity by driving the vector length value closer to 0 for short vectors and closer to 1 for long vectors. The first layer with capsules is called the "Primary Capsules" layer.

$$v_j = \frac{\|s_j\|^2}{1 + \|s_j\|^2} \frac{s_j}{\|s_j\|} \tag{1}$$

as in [3], where s_j represents a given capsule, j, and v_j is the output of capsule j after normalising or *squashing*.

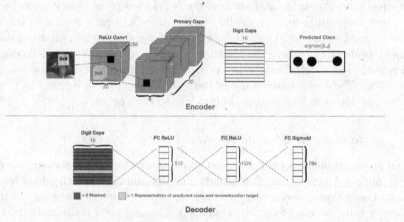

Fig. 1. Architecture of a Capsule Network adapted from [3]

The second capsule layer as in [3] is called "Digit Capsules." It is responsible for the prediction and is expected to have a number of capsules equalling that of the classes in the dataset. Routing between primary capsules and digit capsules is what [3] named "Dynamic Routing" where lower layer capsules *choose* which higher layer capsules to send their information to. In dynamic routing, the lower layer capsules try to predict the output of every higher layer capsule by using a transformation matrix. By multiplying the weights in the transformation matrix with the lower layer capsule output, we get the predicted output of higher layer capsules as in Eq. 2. The predicted output is used to calculate *routing coefficients* c_{ij} which represent the likelihood that a lower layer capsule connects with a higher layer one.

$$\hat{u}_{j|i} = W_{ij} u_i \tag{2}$$

Routing coefficients of a certain layer should sum up to 1, hence c_{ij} is calculated as a softmax of log prior probabilities b_{ij} (Eq. 3). Log priors b_{ij} are initialised to zeros and are learnt over several iterations discriminatively along with other weights in the network and is dependent on the capsules types and locations. They are refined by adding a scalar value representing the *agreement* between two capsules, which can be calculated as a similarity score between the predicted outputs and the activity vector of a higher layer capsule as in Eq. 4. By the end of the training process, the Digit Capsules will contain the instantiation parameters of each class and the probability of the existence of a certain class can be calculated as the magnitude of its corresponding vector. The output vector with highest magnitude is the detected class.

$$c_{ij} = \frac{exp(b_{ij})}{\sum_k exp(b_{ik})} \tag{3}$$

$$b_{ij} = b_{ij} + \hat{u}_{j|i}.v_j \tag{4}$$

The second part of the network is the decoder. In this part, the decoder takes the output vector of the detected class from the encoder while other capsules are masked, and feeds that to a series of fully connected layers to reconstruct the original image. The loss of the network is calculated over two stages: firstly, the Margin Loss is calculated for the encoder, and secondly, a Reconstruction Loss is calculated for the decoder. The margin loss (Eq. 5) is calculated for each output capsule and aims to allow a certain capsule of class k to have a long instantiation vector if and only if the entities of that class exist in the image. The total margin loss is the sum of all digit capsule losses. The reconstruction loss on the other hand is calculated using mean squared error (MSE) between the reconstructed image and the original. The total loss of a Capsule Network is the sum of the margin and the reconstruction losses with a down-weighing factor for the decoder loss (0.0005) so as not to dominate the margin loss.

$$L_k = T_k \max(0, m^+ - \|\mathbf{v}_k\|)^2 + \lambda(1 - T_k) \max(0, \|\mathbf{v}_k\| - m^-)^2 \tag{5}$$

as in [3] where $T_k = 1$, $m^+ = 0.9$, $m^- = 0.1$, $\lambda = 0.5$ is a weighing factor to minimise the effect of loss of absent digits from shrinking the length of other digits' activity vectors.

3   Applications of Capsule Networks on Medical Images

In this section, we will review the latest approaches in the literature on using CapsNets for CADe/x tasks. We organise the literature reviewed by the part of the body and the type of disease on which CapsNets have been used for detection and diagnosis. This is not a comprehensive review, and the techniques reviewed below are only examples representative of CapsNets potential for CADe/x tasks.

3.1   Brain Injuries and Tumours

In a work by Cheng et al. [10], a novel approach for classifying brain tumours in Magnetic Resonance Images (MRIs) was proposed. The authors used a CapsNet for identifying existing tumours and predicting their types from one of three categories: meningioma, glioma, or pituitary tumour. The dataset used contained 3,064 images and is imbalanced. The proposed model accepts two inputs; the first input is a 512×512 full image of a brain MRI scan, and the second input is an image with only the brain tumour region with a size of 128×128. A max-pooling operation was applied to the larger input to reduce its size to 128×128 to decrease the number of parameters in the image. Then, two convolutional layers were applied to each input separately to extract a feature map for each. After that, the resulting feature maps were merged depth-wise and were used as the input for Primary Caps. The output of the network is produced by a "Data capsule" layer which directly succeeds the Primary Caps layer and consists of three capsules representing the three possible classes. Another novelty in this work was the proposal of a modified total loss function where the authors replaced the reconstruction loss with a classification loss for the last layer in the network and calculated that as a cross-entropy loss. The authors argue that the reconstruction loss has no significant effect on the classification, thus they removed it from the total loss function for the model. The reported results of this approach were very good, with an average performance of 93.5% accuracy. The authors compared their model to other architectures they implemented including the original CapsNet as in [3] and a CNN as in [11], and the proposed approach outperformed both. Commenting on the results reported in this work, more metrics were needed to better evaluate the model's performance such as sensitivity, specificity and F1-score, especially since the dataset was imbalanced. Moreover, the authors applied a max-pooling operation on one of the inputs which opposes the concept of CapsNets which refrains from using pooling operations to avoid loss of information.

Afshar et al. [12] started a series of works on using CapsNets for the classification of brain tumours in MRI scans. The dataset used for all works was the same and consisted of 3,064 images of three malignancy categories in addition to the tumour region coordinates for each image. In their first paper [12], they explored different CapsNets architectures that could potentially perform best on the given task. They used the original CapsNet architecture as in [3] but changed some of its hyperparameters. The reported results showed that CapsNets with feature maps reduced from 256 down to 64 resulted in the highest accuracy. They also reported that CapsNets outperform a CNN architecture adopted from [11]. In a later work [13], the authors explored the effect of feeding tumour region coordinates to CapsNets alongside the full brain scan. They were motivated by the hypothesis that CapsNets are sensitive to background noise in images and have the tendency to model everything, and with the high level of detail captured in MRIs, this could impede CapsNets performance. They fed the full brain scans into a CapsNet based on an architecture that they developed in their previous work [12]. The tumour boundary coordinates were concatenated with the unmasked output vector, i.e., the vector of the highest magnitude, and were fed into two fully connected layers. The final layer of the proposed architecture was a softmax layer which output the probability of each class being present. The rationale for concatenating the tumour boundary coordinates with the last capsule layer output was to get the network to pay more attention to the tumour region and not get distracted by extra details in the image. The results for this approach showed an enhanced performance over the authors' previous work [12] with 90.89% accuracy. Additionally, this approach outperformed feeding segmented tumours to the same CapsNet with an increase of 4.33% in accuracy. The results only reported the accuracy of the model with no tracking of other metrics. In a later paper [14], the authors explored the interpretability of CapsNet by examining the learned features and determining whether they are discriminative of the three types of brain tumours using T-distributed Stochastic Neighbour Embedding (t-SNE) visualisation technique as in [15]. The authors' findings were that the CapsNet was able to successfully distinguish between the classes with good separability. However, the authors provided a visualisation of the input features and there was minimum overlap between the data points from different classes suggesting ease of separability even with simpler models. The authors went on to reconstruct the input from the learned features of the predicted class. To understand what features have been learned by the CapsNet the authors repeatedly tweaked the output features in the last capsule layer and reconstructed the input. Then, they tried to relate the tweaked and reconstructed inputs to hand-crafted features that human experts use to characterise a tumour such as tumour size, maximum horizontal and vertical diameter and tumour centre. They found that the features learned by the CapsNet were correlated with the hand-crafted features. This finding suggests that CapsNets provide better interpretability than CNNs and can be used for clinical applications where the workings of a model are as important as the prediction. In their final work in this series [16], the authors proposed a novel architecture of a CapsNet with an internal boosting mechanism. Boosting is a committee-based machine learning technique that starts with a weak learner, i.e., model, and repeatedly trains this model while adding more weight to misclassified samples at each iteration, resulting in a better performing model at the end of the training cycle. The authors built on their previous work [13] and reused the CapsNet architecture; the only difference was the training process as described. The results of this work provided more insight into the model's performance as the authors tracked accuracy, sensitivity, specificity and area under the ROC curve (AUC) for the receiver operating characteristic (ROC) curves for each class. They compared their proposed model to the basic CapsNet as in [3] and the results showed that their model outperformed CapsNet with 92.45% accuracy. BoostCaps had better specificity per-class than CapsNet with an average increase of 4.33% . However, the BoostCaps network was behind in sensitivity and AUC for one class, Glioma, and had the same AUC score as that of CapsNet for the Pituitary class. The authors did not investigate the features learned by BoostCaps and did not comment on whether the ensemble technique they used resulted in better learning of tumour characteristics in the images.

3.2   Ophthalmology

The work by Koresh et al. [17] used a CapsNet with its original architecture as in [3] to classify images of the cornea into one of two categories: noiseless or noisy. The dataset used in this work was captured by an Optical Coherence Tomography (OCT) device which captures a cross section of the cornea for thickness assessment. The dataset was small consisting of only 579 images, and no specific preprocessing was mentioned. The authors tracked the accuracy, specificity, sensitivity, positive predictive value (PPV) and negative predictive value (NPV) of the model and reported that the CapsNet achieved 95.65% accuracy, 96.08% sensitivity, 80% specificity, 99.42% PPV and 36.36% NPV on images belonging to all patients. The hyperparameters for the CapsNet were not given.

In another work by Tsuji et al. [18], a slightly modified CapsNet architecture was used for the classification of diabetic neuropathy as a screening system to avert the danger of vision loss in diabetic patients. The dataset consisted of 84,484 OCT images of four classes that are imbalanced. The images were resized to 512×512 and augmented by shifting by 16 pixels in each direction. The CapsNet architecture the authors proposed added four convolutional layers before the primary caps layer, then a final capsule layer with four capsules representing the four possible classes as output. The metrics tracked were only the loss and accuracy, and the CapsNet was compared to an Inception V3 model as in [19]. The results reported a higher accuracy for the Inception model 99.8% compared with 99.6% for the CapsNet model. The authors argue that the comparison is favourable to CapsNet due to its shallowness compared to the Inception model; however, more data were needed about trainable parameters for both models to support the authors' argument.

On the other hand, Koresh et al. [20] used a modified CapsNet for an image classification followed by an image segmentation task on corneal images from OCT devices. The dataset used in this work contained 579 images. The task was to assess corneal thickness prior to LASIK surgery, and for this task the proposed system needed to use two CapsNets over two stages: the first stage used a Class CapsNet as in [3] to classify noisy corneal images and filter them out, and in the second stage, another CapsNet architecture called SegCaps as in [21,22] with a slight modification was employed to segment the corneal layer in the image for the human expert to measure its thickness. The Class CapsNet scored 96.41%, 96.83%, 83.33%, 99.46%. 45.45%, 98.12% on accuracy, sensitivity, specificity, PPV, NPV and F1-score, respectively. The modified SegCaps had an average Dice coefficient accuracy of 97.17%. In the classification task, the CapsNet outperformed other CNN models, and in the segmentation task, the SegCaps outperformed U-Net which is a CNN-based segmentation model. One statement by the authors that could be argued against is that CapsNet have the ability to "visualise" images even if there is heavy noise in them. This is contrary to repeated findings in the literature that one of the limitations of CapsNets is their attempt to model everything in an image, hence, being affected by any noise.

3.3   Cardiac Diseases

In a work by He et al. [23], a novel capsule-based automated segmentation technique was proposed. The authors used cardiac MRIs to segment the Left Ventricle (LV) which plays a key role in quantifying the volumetric functions of the heart. The dataset they used consisted of 1,720 images and they applied a series of transformations to the images prior to feeding them to the network. The preprocessing transformations start with Fast Fourier Transform (FFT) and inverse FFT followed by Canny edge detection and finally, Hough circle detection and Gaussian kernel. After that, the transformed images are fed to SegCaps as in [22] which is a CapsNet architecture modified for image segmentation tasks. The output of SegCaps undergoes further processing by applying Otsu adaptive thresholding algorithm as in [24] to remove residual noise from the segmentation step. The metrics used to measure the model's performance were Dice coefficient, Jacquard coefficient, Average Symmetric Distance (ADS) and Hausdorff Distance (HD). The authors' proposed approach achieved the best results with 92.2% Dice score, 85.9% Jacquard score, 0.226 mm ASD and 3.541 mm HD and a P value of $< 10^{-6}$ and $< 10^{-13}$ when comparing against other approaches such as SegCaps with preprocessing only and SegCaps with postprocessing only, respectively. The reported results supported by the P values of comparative tests are significant and the authors' approach was successful in segmenting LV in cardiac MRIs; however, as the authors mentioned, the results were not compared with other LV segmentation techniques in the literature. Additionally, the authors did not include clinical measures or detect multiple objects in an image.

Bargsten et al. [25] presented an optimised CapsNet for segmentation of blood vessel lumen and wall in Intravascular Ultrasound (IVUS) images. There were two datasets used for this work: one containing 435 annotated frames with a size of 384×384, and the second dataset contained 77 images with a size of 512×512. The authors preprocessed all the images by resizing them into 256×256 and applying random rotations and flips on-the-fly during training. The authors used a CapsNet architecture called Matwo-Capsnet cited in [21] intended for image segmentation tasks. Matwo-CapsNet hyperparameters were optimised to work on IVUS images, as these kind of images usually contain shadows of artefacts that may obscure the target tissues as well as containing speckle noise which tends to make borders between tissues unclear and hard to detect. The objective function used was margin loss as in [3]. The authors used Dice coefficient and Hausdorff distance for evaluating the model's performance. The network that had the best performance had 420,000 trainable parameters, implemented dual routing with three iterations and applied normalisation for the pose matrix as proposed in [21]. The best performing network scored 73.09% on Dice coefficient for vessel wall and 90.84% on the lumen, which is higher than that of a U-Net model cited in [26], trained on the same data, by 2.83% and 0.27%, respectively. The best model's results on average Hausdorff distance were 0.085 mm on the vessel wall segmentation and 0.022 mm on lumen segmentation. The Hausdorff results were better than those for the U-Net model, which suggests better accuracy for the CapsNet over CNN-based segmentation techniques. The authors also noted that CapsNet models had smaller standard deviations, suggesting better stability across trainings.

3.4   Pulmonary Diseases

In a work by Tiwari et al. [27], CapsNet was integrated with other state-of-the-art CNN architectures such as DenseNet as in [28], VGG16 as in [29], ResNet-50 as in [30] and MobileNet as in [31]. The proposed models were named DenseCapsNet, VGGCapsNet, ResCapsNet and MobileCapsNet and were used to detect

COVID-19 in lung CT scans. The dataset used contained 2,482 images of two balanced classes. The images were resized to 128×128. Each of the CNN architectures was pretrained and the full structure was kept except the last dense layer. The weights of all layers were frozen, then the resulting feature maps were passed to the standard CapsNet architecture by [3]. The authors reported that the basic CapsNet performed poorly on lung CT scans and that the proposed architectures outperformed CapsNet. All models had an accuracy of 99%, sensitivity of 99% and F1-score of 99% except ResCapsNet. In similar work by Yousra et al. [32], an ImageNet-pretrained VGG19 network was fused with a CapsNet to detect COVID-19 in chest X-rays. The dataset contained 3,310 images of three imbalanced classes. The authors resized the images to 224×224, normalised and augmented them by applying random rotation, horizontal flips and scaling. The model's performance was evaluated using accuracy, precision and F1-score on which the model scored 94%, 95% and 94%, respectively.

Karnkawinpong et al. [33] used CapsNet for detecting Tuberculosis (TB) in chest X-rays. The images were resized to 32×32 pixels and randomly rotated by $10°$ in each direction to increase the number of images in the dataset. The model architecture was similar to the basic CapsNet in [3] but with two convolutional layers at the beginning of the network instead of one. The model performed worse (86.86% accuracy) than CNN-based models in comparison such as VGG16 (90.79% accuracy) when tested on images with the same rotation range of the training set. However, the performance of the CNN-based models quickly deteriorated (74.17% accuracy) when tested on images with larger rotations of $30°$. CapsNet, on the other hand, showed better robustness towards image rotation and outperformed models in comparison (80.06% accuracy) by a large margin. This result evidences the ability of CapsNet to provide equivariance better than CNNs. It should be noted that significant information would have been lost when the images were downsized, and this may have limited the performance of the CapsNets in this study.

The work by LaLonde et al. [34] was one of the first and few works to investigate the instantiation parameters learned by CapsNets. Their intuition was to examine the ability of capsules to model specific visual attributes within the learned instantiation parameters. The authors proposed a novel architecture, X-Caps, to classify lung nodules in CT scans into one of six classes based on their visual attributes as described by expert radiologists. The dataset contained 1018 CT scans, no preprocessing was applied to the images but downsizing to 32×32. The number of the output classes was set to six, representing the number of visual attributes needed to characterise a tumour such as sphericity, margin, lobulation, texture, subtlety and spiculation. The authors tried to keep the architecture of X-Caps similar to that of the standard CapsNet in [3] but only introduced a few modifications. Firstly, they modified the dynamic routing algorithm to allow child capsules to route information to all parent capsules instead of letting them "choose" one as in the standard CapsNet. This modification allowed a nodule to score high or low on multiple visual attributes at once. Secondly in a branch, the authors attached a fully connected layer to the class capsules, i.e., visual attributes capsules, which served as supervised labels for the X-Caps output vectors. The output of this fully connected layer were a range of scores (1–5) for each visual attribute with 1 interpreted as low evidence of the corresponding attribute and 5 being high evidence. Finally, the authors reconstructed the input images using the X-Caps output vectors and applying subtle variations each dimension to ensure that the desired attributes are being modeled correctly. The results of their work in terms of accuracy in detecting a tumour was higher than other comparable CNN techniques and the standard CapsNet. More importantly in terms of accuracy for each visual attribute, their work outperformed that of an explainable CNN-based technique, 3D Dual-Path HSCNN as in [35], in every attribute by a margin of >10%. This work was able to provide evidence that CapsNets are more interpretable than their counterparts.

A novel approach was proposed by Afshar et al. [36] for the detection of lung nodule malignancy in CT scans and was called 3D-MCN. The authors used three independent CapsNets and fed each one of them inputs from three different sets of training data at different scales. The output vectors of the three CapsNets were concatenated and fed into a fusion module consisting of fully connected layers. The final output of the proposed model is the probability of a nodule being benign or malignant, a binary classification problem based on the information learnt from the three input images scales. The proposed model was evaluated based on accuracy, specificity, sensitivity and AUC, and was compared to other CNN architectures including a 3D-CNN model that the authors developed to mimic their CapsNet-based model in having three independent networks fused. The accuracy of the 3D-MCN model at the final layer was 93.12%, and sensitivity and specificity were 94.94% and 90%, respectively. Although the proposed model was outperformed by a CNN trained with expert hand-crafted features, the 3D-MCN had the highest sensitivity, which is often more important in medical diagnosis as it means the model rarely misclassifies patients who do have a tumour.

4 Discussion

CapsNets have been evaluated on medical images in a number of works in the literature and the results suggest they can have significant benefits over their CNN counterparts. CapsNets were successfully applied to different imaging modalities with diverse underlying features and quite different tasks. Standard CapsNets as cited in [3] might not be the best architecture to use when working on medical images classification tasks that require preprocessing, for example to remove noise, and the addition of several convolutional layers before the Primary Caps may be beneficial as the results in the literature show. For segmentation tasks, several architectures have been proposed to solve these tasks and have outperformed well-known CNN-based segmentation architectures. We have summarised the reviewed techniques based on CapsNets and their performance compared to CNN-based techniques in Tables 1 and 2 below.

Table 1. Summary of CapsNet-based architectures for classification of medical images and comparison with CNN-based techniques

Technique	Use case	Dataset	Preprocessing	Hyperparameters (CapsNet)	Accuracy	Comparison
Multi-input CapsNet by [10]	Diagnosing brain tumours in MRIs	3,064 images, 3 classes, imbalanced	N/A	Batch size: 30 Learning rate: 0.0001 Routing iterations: 4	**93.5%**	CNN by [11], 72.12% accuracy
CapsNet by [12]	Diagnosing brain tumours in MRIs	3,064 images, 3 classes, imbalanced	N/A	Optimizer: Adam Epochs: 10	**78%**	CNN by [11], 61.97% accuracy
CapsNet by [13]	Diagnosing brain tumours in MRIs	3,064 images + tumour coordinates, 3 classes, imbalanced	N/A	Batch size: 16 Learning rate: 0.01 Routing iteration: 3 Optimizer: Adam Epochs: 50 LR decay: 0.9	**90.89%**	Modified CNN based on [11], 88.33% accuracy
BoostCaps by [16]	Diagnosing brain tumours in MRIs	3,064 images + tumour coordinates, 3 classes, imbalanced	Not detailed, input size = 128×128	Batch size: 16 Routing iterations: 3 Epochs: 100	**92.45%**	–
CapsNet by [17]	Filtering corneal images in OCTs	579 images, 2 classes	N/A	N/A	**95.65%**	CNN (MATLAB/ unspecified), 87.5% accuracy
Modified CapsNet by [18]	Diabetic retinopathy screening	84,484 images, 4 classes imbalanced	Resizing to 512×512, shifting up to 16px all directions	Batch size: 128 Optimizer: Adam Epochs: 50	**99.6%**	Inception V3 as as in [19], **99.8%** accuracy
Mobile-CapsNet by [27]	COVID-19 Detection in CT scans	2,482 images, 2 classes, balanced	Resizing to 128×128	Optimizer: Adam Learning rate: 0.0015 Batch size: 64 Epochs: 100	**99%**	MobileNet as in [31], 98% accuracy
CNN-CapsNet by [32]	COVID-19 Detection in X-rays	3,966 images, 3 classes, balanced	Resizing to 224×224, normalisation, rotation, horizontal flips, scaling	Optimizer: SGD Learning rate: $1e-5$ Batch size: 64 Epochs: 120 Dropout: 0.25	**94%**	–
Modified CapsNet by [33]	TB Detection in X-rays	3,310 images, 2 classes, imbalanced	Resizing to 32×32, rotation $(\pm 10^{\circ})$	Batch size: 64 Epochs: 52 Optimizer: Adam Learning rate: 0.001	**80.06-86.68%**	VGG16 as in [29], 74.17-90.79% accuracy
X-Caps by [34]	Lung Malignancy Detection in CT	1,018 CT scans, 2 classes (tumour = 0/1), 6 visual attributes in second branch	N/A	Batch size: 16 Optimizer: Adam Learning rate: 0.02 LR decay: 0.1	**86.39%** in tumour detection	3D DualPath HSCNN% as in [35], 84.20% accuracy in tumour detection
3D-MCN by [36]	Lung Malignancy Detection in CT	2,283 images, 2 classes	Patch extraction with different scales, normalisation, random flipping	Optimizer: Adam	**93.12%**	ResNet as in [30], 89.90% accuracy

Table 2. Summary of CapsNet-based architectures for segmentation of medical images and comparison with CNN-based techniques

Technique	Use case	Dataset	Preprocessing	Hyperparameters (CapsNet)	Dice score	Comparison
SegCaps by [20]	Corneal image Segmentation in OCTs	579 images, 2 classes	De-noising, grayscale, extracting region-of-interest	N/A	**97.17%**	U-Net as in [26], 96.06% Dice score
Modified CapsNet by [23]	Segmentation of Left Ventricle	1,720 images, 2 classes	Random translation, rotation, scaling, sheer	Optimizer: Adam Epochs: 10,000 Learning rate: 0.001 LR decay: 0.05	**92.2%**	–
Optimised Matwo-CapsNet by [25]	Segmentation of Intravascular Ultrasound images	512 images, 3 classes	Resized to 256×256, random rotation, flipping	Optimizer: Adam Learning rate: 0.001 Epochs: 200	avg **81.96%**	U-Net as in [26], avg 80.42% Dice score

5 Conclusions and Recommendations for Future Work

Capsule networks have been proposed as an improvement over CNNs and with the promise of providing equivariance without the loss of information that might happen during the pooling operation in CNNs. In this work, we reviewed approaches in the literature that utilised CapsNets for disease diagnosis in medical images. CapsNet have continued to show improved results over CNNs and outperformed more complex and deeper architectures with significantly smaller datasets. Moreover, CapsNets are showing a potential for increased model interpretability due to the learned instantiation parameters embedded in the capsules. Both of the mentioned characteristics of CapsNets—being able to learn from small datasets and their interpretability—are important for models operating in the medical field, and we can conclude that CapsNets could be more suitable than CNNs for being deployed in CADe/x systems. There are a few limitations for CapsNets: a) they are very computationally expensive to train and, in some cases, to run, which might hinder their deployment on conventional machines found in medical institutions; and b) they require some image preprocessing to remove noise in the input images due to their tendency to model everything which might impact their performance. We recommend more work on improving CapsNet architecture, more efficient architectures are needed that have far less trainable parameters without impact on the performance. This will help with deploying CapsNets on various machines and not being limited to ones with powerful hardware. Finally, we recommend more efforts are directed towards CaspNets interpretability and investigating what these networks learn and how we can seize this potential into advancing the eXplainable Artificial Intelligence (XAI) and Trustworthy AI fields.

References

1. Doi, K.: Computer-aided diagnosis in medical imaging: historical review, current status and future potential. Comput. Med. Imaging Graph. **31**(4), 198–211 (2007). https://doi.org/10.1016/j.compmedimag.2007.02.002
2. Byrne, M.F., et al: Real-time differentiation of adenomatous and hyperplastic diminutive colorectal polyps during analysis of unaltered videos of standard colonoscopy using a deep learning model. Gut **68**, 94–100 (2019). https://doi.org/10.1136/gutjnl-2017-314547
3. Sabour, S., Frosst, N., Hinton, G.E.: Dynamic routing between capsules. In: Advances in Neural Information Processing Systems, pp. 3857–3867. Neural information processing systems foundation, 2017-December (2017)
4. Geirhos, R., Michaelis, C., Wichmann, F.A., Rubisch, P., Bethge, M., Brendel, W.: ImageNet-trained CNNs are biased towards texture; increasing shape bias improves accuracy and robustness. In: 7th International Conference on Learning Representations, ICLR 2019. ICLR (2019). https://doi.org/10.1136/gutjnl-2017-314547
5. Nguyen, H.P., Ribeiro, B.: Advanced Capsule Networks via Context Awareness. In: Tetko, I.V., Kůrková, V., Karpov, P., Theis, F. (eds.) ICANN 2019. LNCS, vol. 11727, pp. 166–177. Springer, Cham (2019). https://doi.org/10.1007/978-3-030-30487-4_14
6. Rudin, C.: Stop explaining black box machine learning models for high stakes decisions and use interpretable models instead. Nat. Mach. Intell. **1**(5), 206–215 (2019). https://doi.org/10.1038/s42256-019-0048-x
7. London, A.J.: Artificial intelligence and black-box medical decisions: accuracy versus explainability. Hastings Center Rep. **49**, 15–21 (2019). https://doi.org/10.1002/hast.973
8. Vaswani, A., et al.: Attention is all you need. In: Advances in Neural Information Processing Systems 2017-December, pp. 5999–6009. Neural information processing systems foundation (2017)
9. Abnar, S.: From Attention in Transformers to Dynamic Routing in Capsule Nets. Samira Abnar (blog). https://samiraabnar.github.io/articles/2019-03/capsule. Accessed 30 Aug 2022
10. Cheng, Y., Qin, G., Zhao, R., Liang, Y., Sun, M.: ConvCaps: multi-input capsule network for brain tumor classification. In: Gedeon, T., Wong, K. W., Lee, M. (eds.) Neural Information Processing. ICONIP 2019. Lecture Notes in Computer Science 11953, pp. 524–534. Springer International Publishing (2019). https://doi.org/10.1007/978-3-030-36708-4_43
11. Paul, J.S., Plassard, A.J., Landman, B.A., Fabbri, D.: Deep learning for brain tumor classification. In: Andrzej, K., Barjor, G. (eds.) Progress in Biomedical Optics and Imaging - Proceedings of SPIE 10137, pp. 1013710. SPIE (2017). https://doi.org/10.1117/12.2254195
12. Afshar, P., Mohammadi, A., Plataniotis, K.N.: Brain tumor type classification via capsule networks. In: 25th IEEE International Conference on Image Processing - ICIP 2018. ICIP 2018, pp. 3129–3133. IEEE Computer Society (2018). https://doi.org/10.1109/ICIP.2018.8451379
13. Afshar, P., Plataniotis, K.N., Mohammadi, A.: Capsule networks for brain tumor classification based on MRI images and coarse tumor boundaries. In: 44th IEEE International Conference on Acoustics, Speech and Signal Processing - ICASSP 2019. ICASSP 2019. 2019-May, pp. 1368–1372. Institute of Electrical and Electronics Engineers Inc. (2019). https://doi.org/10.1109/ICASSP.2019.8683759

14. Afshar, P., Plataniotis, K.N., Mohammadi, A.: Capsule networks' interpretability for brain tumor classification via radiomics analyses. In: 26th IEEE International Conference on Image Processing - ICIP 2019. ICIP 2019. 2019-September, pp. 3816–3820. IEEE Computer Society (2019). https://doi.org/10.1109/ICIP.2019. 8803615

15. Van Der Maaten, L., Hinton, G.: Visualizing data using t-SNE. J. Mach. Learn. Res. **9**, 2579–2625 (2008)

16. Afshar, P., Plataniotis, K.N., Mohammadi, A.: BoostCaps: a boosted capsule network for brain tumor classification. In: 42nd Annual International Conferences of the IEEE Engineering in Medicine and Biology Society - EMBC 2020. EMBC 2020. 2020-July, pp. 1075–107. Institute of Electrical and Electronics Engineers Inc. (2020). https://doi.org/10.1109/EMBC44109.2020.9175922

17. Koresh, H.J.D., Chacko, S.: Classification of noiseless corneal image using capsule networks. Soft Comput. **24**, 16201–16211 (2020). https://doi.org/10.1007/s00500-020-04933-5

18. Tsuji, T., et al.: Classification of optical coherence tomography images using a capsule network. BMC Ophthalmol. **20**, 114 (2020). https://doi.org/10.1186/s12886-020-01382-4

19. Szegedy, C., Vanhoucke, V., Ioffe, S., Shlens, J., Wojna, Z.: Rethinking the inception architecture for computer vision. In: 29th IEEE Conference on Computer Vision and Pattern Recognition - CVPR 2016. CVPR 2016. 2016-December, pp. 2818–2826. IEEE Computer Society (2016). https://doi.org/10.1109/CVPR.2016. 308

20. Koresh, H.J.D., Chacko, S., Periyanayagi, M.: A modified capsule network algorithm for oct corneal image segmentation. Patt. Recogn. Lett. **143**, 104–112 (2021). https://doi.org/10.1016/j.patrec.2021.01.005

21. Bonheur, S., Štern, D., Payer, C., Pienn, M., Olschewski, H., Urschler, M.: Matwo-CapsNet: a multi-label semantic segmentation capsules network. In: Shen, D., et al. (eds.) MICCAI 2019. LNCS, vol. 11768, pp. 664–672. Springer, Cham (2019). https://doi.org/10.1007/978-3-030-32254-0_74

22. LaLonde, R., Xu, Z., Irmakci, I., Jain, S., Bagci, U.: Capsules for biomedical image segmentation. Med. Image Anal. **68**, 101889 (2021). https://doi.org/10.1016/j.media.2020.101889

23. He, Y., et al.: Automatic left ventricle segmentation from cardiac magnetic resonance images using a capsule network. J. X-Ray Sci. Technol. **28**, 541–553 (2020). https://doi.org/10.3233/XST-190621

24. Otsu, N.: A threshold selection method from gray-level histograms. IEEE Trans. Syst. Man Cybern. **9**, 62–66 (1979). https://doi.org/10.1109/TSMC.1979.4310076

25. Bargsten, L., Raschka, S., Schlaefer, A.: Capsule networks for segmentation of small intravascular ultrasound image datasets. Int. J. CARS **16**(8), 1243–1254 (2021). https://doi.org/10.1007/s11548-021-02417-x

26. Hssayeni, M.D., Croock, M.S., Salman, A.D., Al-khafaji, H.F., Yahya, Z.A., Ghoraani, B.: Intracranial hemorrhage segmentation using a deep convolutional model. Data **5**, 14 (2020). https://doi.org/10.3390/data5010014

27. Tiwari, S., Jain, A.: A lightweight capsule network architecture for detection of COVID-19 from lung CT scans. Int. J. Imaging Syst. Technol. **32**(2), 419–434 (2022). https://doi.org/10.1002/ima.22706

28. Huang, G., Liu, Z., Van Der Maaten, L., Weinberger, K.Q.: Densely connected convolutional networks. In: 30th IEEE Conference on Computer Vision and Pattern Recognition - CVPR 2017. CVPR 2017. 2017-January, pp. 2261–2269 (2017). https://doi.org/10.1109/CVPR.2017.243

29. Simonyan, K., Zisserman, A.: Very deep convolutional networks for large-scale image recognition. In: 3rd International Conference on Learning Representations - ICLR 2015. ICLR 2015 (2015)

30. He, K., Zhang, X., Ren, S., Sun, J.: Deep residual learning for image recognition. In: 29th IEEE Conference on Computer Vision and Pattern Recognition - CVPR 2016. CVPR 2016. 2016-December, pp. 770–778 (2016). https://doi.org/10.1109/CVPR.2016.90

31. Srinivasu, P.N., Sivasai, J.G., Ijaz, M.F., Bhoi, A.K., Kim, W., King, J.J.: Classification of skin disease using deep learning neural networks with MobileNet V2 and LSTM. Sensors **21**(8), 2852 (2021). https://doi.org/10.3390/s21082852

32. Yousra, D., Abdelhakim, A.B., Mohamed, B.A.: A novel model for detection and classification coronavirus (COVID-19) based on Chest X-Ray images using CNN-CapsNet. In: Corchado, J.M., Trabelsi, S. (eds.) Sustainable Smart Cities and Territories. SSCTIC 2021. Lecture Notes in Networks and Systems 253, pp. 187–199 (2021). https://doi.org/10.1007/978-3-030-78901-5_17

33. Karnkawinpong, T., Limpiyakorn, Y.: Chest X-Ray analysis of tuberculosis by convolutional neural networks with affine transforms. In: 2nd International Conference on Computer Science and Artificial Intelligence - CSAI 2018. CSAI 2018, pp. 90–93. Association for Computing Machinery (2018). https://doi.org/10.1145/3297156.3297251

34. LaLonde, R., Torigian, D., Bagci, U.: Encoding visual attributes in capsules for explainable medical diagnoses. In: 23rd International Conference on Medical Image Computing and Computer Assisted Intervention - MICCAI 2020. MICCAI 2020. Lecture Notes in Computer Science 12261, pp. 294–304 (2020). https://doi.org/10.1007/978-3-030-59710-8_29

35. Shen, S., Han, S.X., Aberle, D.R., Bui, A.A., Hsu, W.: An interpretable deep hierarchical semantic convolutional neural network for lung nodule malignancy classification. Expert Syst. Appl. **128**, 84–95 (2019). https://doi.org/10.1016/j.eswa.2019.01.048

36. Afshar, P., et al.: A 3D multi-scale capsule network for lung nodule malignancy prediction. Sci. Rep. **10**, 7948 (2020). https://doi.org/10.1038/s41598-020-64824-5

Introducing an Atypical Loss: A Perceptual Metric Learning for Image Pairing

Mohamed Dahmane[✉][iD]

CRIM- Computer Research Institute of Montreal, Montreal, QC H3N 1M3, Canada
mohamed.dahmane@crim.ca
http://www.crim.ca/en

Abstract. Recent works have shown an interest in comparing visually similar but semantically different instances. The paired Totally Looks Like (TLL) image dataset is a good example of visually similar paired images to figure out how humans compare images. In this research, we consider these more generic annotated categories to build a semantic manifold distance. We introduce an atypical triplet-loss using the inverse Kullback-Leibler divergence to model the distribution of the anchor-positive (a-p) distances. In the new redefinition of triplet-loss, the anchor-negative (a-n) loss is conditional to the a-p distance distribution which prevents the loss correction fluctuations in the plain summed triplet-loss function of absolute distances. The proposed atypical triplet-loss builds a manifold from relative distances to a "super" anchor represented by the a-p distribution.

The evaluation on the paired images of the TLL dataset showed that the retrieving score from the first candidate guess (top-1) is 75% which is ×2.5 higher compared to the recall score of the baseline triplet-loss which is limited to 29%, and with a top-5 pairing score as high as 78% which represents a gain of ×1.4.

Keywords: Atypical loss · Metric learning · Visual relationship · Image pairing · Image perception · Image retrieval

1 Introduction

Human reasoning about objects and their similarity is classified depending on the object proximity in the representational geometry. It depends also on their features; similar objects have more shared features. Similarity can also be measured in a structural space by matching the structural elements of the compared objects. The similarity between objects has been also assimilated to the length of the sequence of transformations required to transform one object into another [8].

To perceive objects human build conceptual representational models. In semantic vision, deep neural networks allow to elaborate representational models to apprehend visual objects [12]. There are significant objects that have the same visual aspect but which do not belong to the same semantic classes. Relational

N. El Gayar et al. (Eds.): ANNPR 2022, LNAI 13739, pp. 81–94, 2023.
https://doi.org/10.1007/978-3-031-20650-4_7

reasoning is studied in psychological research and cognitive assessments to establish a relational order of similarity in these cases [14]. Computer vision algorithms have been elaborated to evaluate the ability to classify different objects across their semantically meaningful class boundaries [5]. The investigated approaches permit to understand how human perception links shape and color processing to build the similarity reasoning process [23]. Deep Neural Networks (DNNs) have proven their ability in classifying categorical objects that belong to the same class hierarchy with visual similarity however there is still an unaccountable aspect in assessing semantic similarity. The existing deep models classify with high level of performance some specific categories of objects but are not good for generic categories. Semantic vision leans to more generic categories. Most of the proposed approaches in the literature consider feature generated by pre-trained CNNs to gather a set of visual descriptive attributes from different levels of abstraction such as the shape of the object, its texture and color. Since these are biased by the object category, an adequate learning should build similarity function that evaluates how similar or related are the compared objects.

Deep metric learning models are more dedicated to address both similarity and relation. In this article, we define an atypical triplet-loss learning based on the Kullback-Leibler divergence that constrains the positive to anchor distance distribution, whereas the negative to anchor distances are learned as a distance maximization.

2   Related Work

To learn similarity between images, several proposed methods present the problem as a metric learning problem [15] or semantic graph construction to define a distance based image metric classification [6]. The authors in [4] proposed a graph-based image representation denoting the relationship between the image regions. Siamese neural networks (SNN) as coupled neural networks were introduced for distance metric learning [1,22]. They compare feature pairs to establish their semantic similarity. The training loss aims to regroup the embeddings of the identical samples in the same region and move away from this region those of dissimilar samples. Sanders et al. in [21] presented a CNN-predicted representation with a formal psychological model to predict human categorization behavior. In a similar way, the authors in [18] used a SNN to evaluate image similarity on the Totally Looks Like dataset.

Rosenfeld et al. [19] build from a popular entertainment website[1], the TTL dataset of images that human have paired as being similar or not. The benchmark dataset is adopted in neuro-cognitive and computer vision research to figure out and simulate the human behavior by retrieving the various factors that are relevant to the image to image pairing task [16]. The human visual system uses similarity in many tasks; addressing this main component is important to push the understanding of the visual human perception beyond the simple tasks of object detection and scene recognition.

---

[1] https://memebase.cheezburger.com/totallylookslike.

Fig. 1. Samples of paired triplet images –anchor, positive, negative– from the TTL dataset.

The collected image pairs (Fig. 1) reflect the features that human use in guessing for similarity which include global scene feature matching, face resemblance, shape and texture similarity, color matching etc. Some of these features are inherently explicit but others are inferred.

Visual attributes which are defined as semantically understandable by human and visually detectable by machine [7] play an important role in high level image understanding, in interactive image search [9,13], and in transfer learning as analogous attributes [2].

Despite the great advances of the last decade in representation learning, comparing class-unrelated images still presents a great challenge in computer vision due to the high dimensional search and mapping space. The subjectivity of the task itself adds yet another difficulty in similarity assessment. When it comes to compare abstract concept-related/unrelated images most of the proposed metrics are less effective.

Some of the research works adopted deep features with standard metrics [25]. Others suggested a learning of an adaptation function [18] which consists in tacking the average of two directed losses on feature representation from a frozen pre-trained CNN model. In the same way, the authors in [19] measured distances between embedding extracted from general deep CNN models trained on image classification and face identification. Perceptual similarity can be directly learned in an Euclidean space by comparing distances. In this category, triplet-networks [11] as Siamese networks represent good examples of models that can learn useful representations from different manifolds [20].

3   Learning Atypical Perceptual Similarity

Siamese networks are basically trained with contrastive loss function as illustrated in Eq. (1) where D refers to the distance between the image representations x_i and x_j, and m stands for a margin separating the positive and negative samples which are formed by pairs of images. The positive pairs are from the same classes whereas samples containing pairs from different classes are labelled as negative samples. The contrastive objective/loss function aims to push away the negative pairs and bring closer the positives ones [3].

$$\mathcal{L} = \begin{cases} D(x_i, x_j) & \text{if label}(i) = \text{label}(j) \\ \max(0\,,\, m - D(x_i, x_j)) & \text{otherwise.} \end{cases} \tag{1}$$

With the triplet-loss training, the sampling includes anchors with a level of relativity compared to the contrastive loss. Each sample of the triplet-loss combines anchor, positive, and negative examples. The objective function enforces the proximity of the positive and the anchor example, and at the same time penalizes the proximity between the anchor and the negative examples. Figure 2 illustrates the different components of the baseline triplet-network where x_{anc}, x_{pos}, and x_{neg} represent the encoding of the anchor, positive and negative images respectively. Equation (2) formalizes the triplet-loss objective with a margin m.

$$\mathcal{L} = \max\left(D(x_{anc}, x_{pos}) - D(x_{anc}, x_{neg}) + m\,,\, 0\right) \tag{2}$$

3.1   The Baseline Triplet-Network

The task at hand is formulated as a retrieval task since we want to evaluate the learned visual attributes and how they connect to the human pairing. The TLL dataset was designed to find evidence that connects some aspects of the machine image retrieval with the human image pairing. The paired images of the TTL data are grouped into two subsets namely left and right subsets. Given a left anchor image, the purpose is to retrieve the most similar positive match from the right subset.

We trained a triplet-network with different backbones. A reference implementation of the baseline model is available from the Keras repository[2].

---

[2] https://github.com/keras-team/keras-io/blob/master/examples/vision/siamese_network.py.

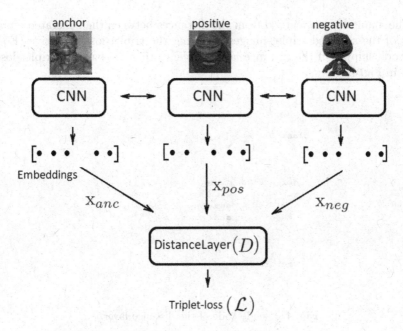

Fig. 2. The architecture of the baseline triplet-network.

The Tensorflow-like pseudo code in Fig. 3 illustrates the baseline model built on the top of a residual network model (RestNet50) which was trained on ImageNet. To learn a representation from the embeddings of the triplet images, a sub-network is appended at the bottom of the model alternating dense and batch normalization layers.

```
def Triplet_network():

    # A code snippet of the baseline
    #  Triplet_network using ResNet50

    base_cnn = resnet.ResNet50()
    f1 = Flatten()(base_cnn.output)
    d1 = Dense(512, activ="relu")(f1)
    d1 = BatchNormalization()(d1)
    d2 = Dense(256, activ="relu")(d1)
    d2 = BatchNormalization()(d2)
    out = Dense(256)(d2)

    # output the embeddings
    embedding = Model(base_cnn.input, out)
    return embedding
```

Fig. 3. Pseudo code of the baseline triplet-network.

The manifolds are learned from the distances between the generated embeddings of the sampled triplet-images following the triplet-loss objective (Eq. 2). The code snippet of the distance layer implementing the baseline triplet-loss is given in Fig. 4.

```
class DistanceLayer(layers.Layer):

    # The baseline distance layer

    def call(self, anc, pos, neg):

        ap_dist = sum((anc - pos)^2)
        an_dist = sum((anc - neg)^2)

        # return the distance  tuple
        #  (dist(anc,pos), ·dist(anc,neg))
        return  (ap_dist, an_dist)
```

Fig. 4. Pseudo code of the distance layer.

As we are interested in the similarity comparison, in what follows the focus will be particularly put on the distance layer of the triplet-network rather on the model itself.

3.2   The Atypical Perceptual Similarity

In this work, we propose an atypical loss function that defines a perceptual similarity. The objective is to learn to minimize the similarity comparison between anchor images and positive images, and maximize the dissimilarity between anchor images and negative images.

The conventional baseline distance layer is defined as a sum of absolute Euclidean distances as shown in Fig. 4. An adaptation function learning module was proposed in [18] as a redefinition of the distance layer that turns the distance metric to an average of two directed (left-right and right-left) cross-entropy losses computed using a cosine similarity metric as shown in the code snippet of Fig. 5. It represents a negative log probability of the positive loss unless an asymmetric similarity metric is used. The inputs of the loss represent the outputs of the adaptation layer which consists of a dense layer referred as a learnable multiplication matrix, and a relu activation layer. The inputs of the adaptation layer are obtained as an embedding of the input image from a frozen backend network.

The atypical loss module, we proposed in this work aims to avoid seesawing side effect of the loss function as a sum of two similar left-to-right and right-to-left loss functions. Also, we guess that the effect of the order in the adaptation

```
def Adaptation_learning_module():

    #A code snippet of the Adaptation
    # function learning module

    loss_LR = CrossEntropy(
            Softmax(< a_L, a_R>*σ),1)

    loss_RL = CrossEntropy(
            Softmax(< a_R, a_L>*σ),1)

    loss = 0.5*(loss_LR + loss_RL)

    return loss
```

Fig. 5. Pseudo code of the adaptation learning module from [18]. The function $< . >$ refers to the cosine similarity. L and R refer to the left and right images.

function may be attenuated by a random pairing direction sampling of the input image pairs.

The baseline triplet-loss distance layer (Fig. 4) that utilizes anchor samples relies on absolute loss distances, and the adaptation function learning module of Fig. 5 defines a classification loss function which compares embeddings. The proposed atypical triplet-loss defines relative distances. It is an atypical loss function of positive and negative losses denoted respectively as ap_loss and an_loss in Fig. 6. The positive loss is computed as a Kullback-Leibler divergence to impose predefined distribution to the anchor-to-positive distance. We will consider a zero-centered distribution with a unit standard deviation as it is more convenient. On the other hand, the negative loss an_loss forces the model to push the negative examples away from the center of the distribution of the positive samples. The centered distribution can be seen as a good estimate of a zero-centered "super" anchor to which all the similarities are relatively assessed.

Constraining the distribution of positive distances to $\mathcal{N}(\mu, \text{var})$ leads to a loss objective $\mathcal{L}_{ap}$ as a function of the anchor-positive distances as in Eq. 3.

$$\begin{aligned}\mathcal{L}_{ap} &= KL_{div}\big(\mathcal{N}(\mu, \text{var}), \mathcal{N}(\mathbf{0}, \mathbf{1})\big) \\ &= 0.5\big(-\log(\text{var}) - 1 + \text{var} + \mu\big)\end{aligned} \tag{3}$$

In the other hand the negative loss objective $\mathcal{L}_{an}$ tries to penalize the anchor-negative distances that are getting closer to the distribution center $\mu_{ap} = 0$. So having $D(\mathrm{x}_{anc}, \mathrm{x}_{pos}) = 0$, Eq. 4 can be easily derived from Eq. 2.

$$\mathcal{L}_{an} = \max\big(-D(\mathrm{x}_{anc}, \mathrm{x}_{neg}) + m, 0\big) \tag{4}$$

Finally, the total atypical triplet-loss is given in (5).

$$\mathcal{L} = \mathcal{L}_{ap} + \mathcal{L}_{an} \tag{5}$$

```
def Atypical_Learning_Module():
    # A code snippet of the proposed
    #   Atypical function learning module

    ap_dist = anchor - positive
    an_dist = anchor - negative

    μ = Identity(ap_dist)
    log_var = Dense(ap_dist)

    ap_loss = -0.5 * (1 + log_var
                - μ^2 - exp(log_var))

    an_loss = max(-an_dist + m, 0)
    loss = ap_loss + an_loss

    return loss
```

Fig. 6. Pseudo code of the Atypical Learning Module.

4 Experimentation

4.1 The TTL Benchmark

To evaluate our perceptual loss, a series of experiments were carried out on the Totally Looks Like dataset that contains 6016 paired images from the wild. The consistency of the data was verified with additional human experiments [19]. The context is to build a machine-based representation to pair images by learning relevant features that human naturally use in assessing similarity.

We followed the same evaluation protocol as in [18] by considering only the TTL_{obj} subset which considers pairs of images that do not include easy detectable face pairs where faces are present in both left and right images. The left out subset is referred as TTL_{faces}. The MTCNN [24] face detector, a deep cascaded multitask face detector framework, was used to detect pairs of faces. Faces with a higher detection confidence were removed ant not considered in TTL_{obj}. We end up with 5523 image pairs with at most one face image per sample. The subset was divided in a ratio of 75% for train and 25% for test. The exploration considers the left to right and also the right to left evaluation.

The margin m of the Atypical Learning Module was fixed to $m = 3$ in Fig. 6. Empirically, for a normal distribution the three sigma rule guarantees that almost all values belong to the interval within three standard deviations of the mean.

The 256-dimensional image embedding of the triplet-network (Fig. 3) was the input of the conventional Distance Layer of Fig. 4, and the input of our Atypical Learning Module (Fig. 6). The training cycle was fixed to 1000 epochs without using any early stopping criteria as there is no validation set. However, due to the limited data we adopted a sampling method to evaluate the baseline model and the proposed module. Ten trial runs were performed, at each run the data were randomly split in 75% for train and 25% for test. The final recall scores are averaged over the ten runs.

4.2   Evaluation

The method was evaluated on some well known base models with demonstrated upset potential in visual classification such as residual networks [10]. We also tested the learning module with the CLIP visual models [17] that were trained on 400M Internet images in a related task of text and image pairing. We focused on both ResNet and Vision Transformer-based architecture from the CLIP's series, in particular, the residual network RN50x4 that provided interesting improvements in assessing visual similarity of images [18].

In Table 1 and 2, a comparison between our atypical learning module and the baseline triplet-loss function is reported along with the performance of the adaptation learning module from [18]. The left to right top-1 recall scores are presented in Table 1. Note that the training was only done by left to right pairing, i.e., using only anchor to positive pairing direction. In Table 2, all the models were evaluated using the reverse right to left pairing (i.e., positive to anchor). The optimistic asymmetrical recall that considers the best prediction from the two possible pairing directions was omitted as it is less prevalent in image querying.

Table 1. Top-1 recall scores- Evaluation of Left to Right pairs.

| Base model | Learning function | | |
	Triplet loss[a] (baseline)	Atypical triplet-loss[a] (our)	Adaptation loss[b] ([18])
ResNet50	0.1737 ± 0.0209	0.7256 ± 0.0039	0.1469 ± 0.0047
ResNet101	0.2098 ± 0.0258	0.7347 ± 0.0028	0.1913 ± 0.0034
RN50x4	0.3993 ± 0.0097	$\mathbf{0.7489 \pm 0.0038}$	0.3939 ± 0.0071
ViT-B/32	0.2915 ± 0.0137	$\mathbf{0.7490 \pm 0.0055}$	0.3540 ± 0.0046

[a]Trained and tested only on L-R pairs.

[b]Trained and tested on both L-R and R-L pairs.

The proposed atypical loss gave almost similar scores whether conventional left to right or right to left assessment was considered. Although overall a slight increase in performance is recorded for the atypical loss with the left to right pairs as they were used for training. However, the gain was more important for the baseline triplet loss. The atypical loss learning module was able to learn the asymmetric matching even if the training was performed in a one way (i.e., left to right) pairing.

The RN50x4 and ViT-B/32 from the CLIP series achieved the best top-1 recall scores of 74.9% which is more than twice the score of the baseline model that was limited to an average score of 29.1%. It is also well above the 39.9% top-1 recall score of the adaptation-loss.

The atypical loss showed from both left to right and right to left evaluations almost similar scores despite the training was performed on the left to right image pairs. Unlike the baseline triplet loss that is sensitive to the pairing direction. Moreover, the atypical loss exhibited a reduced prediction error variability in

Table 2. Top-1 recall scores- Evaluation of Right to Left pairs.

| Backbone model | Learning function | |
	Triplet loss (baseline)	*Atypical triplet-loss (our)*
ResNet50	0.1547 ± 0.0183	0.7250 ± 0.0043
ResNet101	0.1743 ± 0.0163	0.7355 ± 0.0029
RN50x4	0.3102 ± 0.0079	**0.7513 ± 0.0038**
ViT-B/32	0.2377 ± 0.0256	**0.7508 ± 0.0052**

the same range as the adaptation loss whereas the average prediction errors were more pronounced for the baseline triplet loss.

Figure 7 and Fig. 8 show the top-5 most similar retrieved images given a query anchor image. The retrieving ability in Fig. 7 was somewhat limited with only one valid guess exhibiting a limited ability in perceptual similarity assessment of the models trained with the baseline triplet-loss. However, all the models trained with the atypical learning module reported a much better top-5 recall scores. The top-5 predictions obtained by ViT-B/32 trained with the atypical triplet loss are depicted in Fig. 8 with a higher rate of retrieved positive images.

Fig. 7. Top-5 predictions from ViT-B/32 trained with the baseline triplet-loss.

Figure 9 and Fig. 10 portray the averaged conventional left to rigt and right to left top-k recall score distribution with k ranging from 1 to 100. The dashed pattern lines correspond to the scores of the baseline triplet-loss whereas the plain lines represent the scores of the atypical learning function used to train the

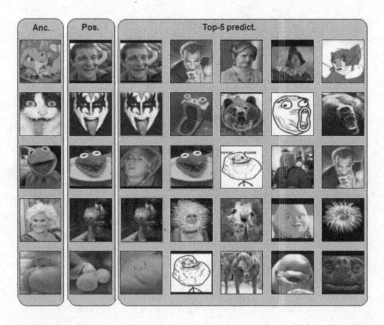

Fig. 8. Top-5 predictions from ViT-B/32 trained with our atypical triplet-loss.

respective model. Obviously, the proposed learning module has made it possible to make a significant performance gain in the top-5 predictions maintaining a higher tendency.

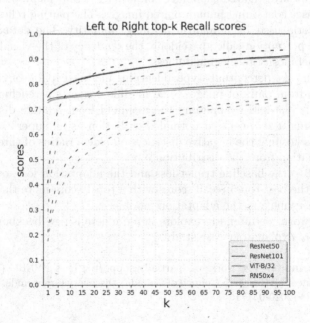

Fig. 9. Comparison of the top-k recall scores using left to right pairing.

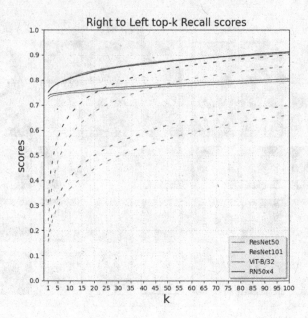

Fig. 10. Comparison of the top-k recall scores using right to left pairing.

5   Conclusion

In this work, we investigated the image similarity assessment problem. We used the TTL dataset containing human paired images. The pairing relies on certain identified criteria used by human in defining similarity. The dataset recalls a form of visual perception called pareidolia, the tendency of the visual perception to associate nebulosity to meaningfulness.

We introduced a perceptual atypical learning module capable of visually linking semantics from pairs of images. The module suggests a redefinition of the conventional triplet-loss by building the manifold from relative distance comparison by using a zero-centred normal distribution as a "super" anchor. The learning forces keeping the negative distances outside a radius of three times the standard deviation from the distribution mode.

Compared to the baseline triplet-loss and the adaptation loss, our atypical-loss achieved the best top-5 recall scores with a top-1 score more than twice the score of the baseline and the adaptation module.

As future work, we intend to explore in more details new backbone architectures in a more explanatory case study.

Acknowledgements. The work was partially supported by a NSERC (Natural Sciences and Engineering Research Council of Canada) Discovery Grant under grant agreement No RGPIN-2020-05171.

We gratefully acknowledge the support of the Computer Research Institute of Montreal (CRIM), the Ministère de l'Économie et de l'Innovation (MEI) of Quebec, and The Natural Sciences and Engineering Research Council of Canada (NSERC).

References

1. Bromley, J., Guyon, I., LeCun, Y., Säckinger, E., Shah, R.: Signature verification using a "Siamese" time delay neural network. In: Cowan, J., Tesauro, G., Alspector, J. (eds.) Advances in Neural Information Processing Systems, vol. 6. Morgan-Kaufmann (1993)
2. Chen, C.-Y., Grauman, K.: Inferring analogous attributes. In: 2014 IEEE Conference on Computer Vision and Pattern Recognition, pp. 200–207 (2014). https://doi.org/10.1109/CVPR.2014.33
3. Chopra, S., Hadsell, R., LeCun, Y.: Learning a similarity metric discriminatively, with application to face verification. In: 2005 IEEE Computer Society Conference on Computer Vision and Pattern Recognition (CVPR 2005), vol. 1, pp. 539–546 (2005). https://doi.org/10.1109/CVPR.2005.202
4. de Mauro, C., Diligenti, M., Gori, M., Maggini, M.: Similarity learning for graph-based image representations. Patt. Recogn. Lett. **24**(8), 1115–1122 (2003)
5. Deselaers, T., Ferrari, V.: Visual and semantic similarity in ImageNet. In: IEEE Conference on Computer Vision and Pattern Recognition (CVPR), pp. 1777–1784 (2011)
6. Fang, C., Torresani, L.: Measuring image distances via embedding in a semantic manifold. In: Fitzgibbon, A., Lazebnik, S., Perona, P., Sato, Y., Schmid, C. (eds.) ECCV 2012. LNCS, vol. 7575, pp. 402–415. Springer, Heidelberg (2012). https://doi.org/10.1007/978-3-642-33765-9_29
7. Feris, R.S., Lampert, C., Parikh, D.: Visual Attributes, 1st edn. Springer Publishing Company, Incorporated (2018)
8. Goldstone, R., Son, J.: Similarity: the oxford handbook of thinking and reasoning. In: Holyoak, K.J., Morrison, R.G. (eds.) November 2012
9. Gordo, A., Almazan, J., Revaud, J., Larlus, D.: End-to-end learning of deep visual representations for image retrieval. Int. J. Comput. Vis. **124**, 237–254 (2017). https://doi.org/10.1007/s11263-017-1016-8
10. He, K., Zhang, X., Ren, S., Sun, J.: Deep residual learning for image recognition (2015). https://doi.org/10.48550/ARXIV.1512.03385, https://arxiv.org/abs/1512.03385
11. Hoffer, E., Ailon, N.: Deep metric learning using triplet network. In: Feragen, A., Pelillo, M., Loog, M. (eds.) SIMBAD 2015. LNCS, vol. 9370, pp. 84–92. Springer, Cham (2015). https://doi.org/10.1007/978-3-319-24261-3_7
12. Jozwik, K.M., Kriegeskorte, N., Storrs, K.R., Mur, M.: Deep convolutional neural networks outperform feature-based but not categorical models in explaining object similarity judgments. Front. Psychol. **8**, 1726 (2017)
13. Kovashka, A., Parikh, D., Grauman, K.: WhittleSearch: image search with relative attribute feedback. In: 2012 IEEE Conference on Computer Vision and Pattern Recognition, vol. 115, pp. 2973–2980, May 2012
14. Krawczyk, D.C., et al.: Distraction during relational reasoning: the role of prefrontal cortex in interference control. Neuropsychologia **46**(7), 2020–2032 (2008)
15. Kulis, B.: Metric learning: a survey. Found. Trends Mach. Learn. **5**(4), 287–364 (2013)

16. Lindsay, G.W.: Convolutional neural networks as a model of the visual system: past, present, and future. J. Cogn. Neurosci. **33**(10), 2017–2031 (2021)
17. Radford, A., et al.: Learning transferable visual models from natural language supervision. In: ICML, pp. 8748–8763 (2021)
18. Risser-Maroix, O., Kurtz, C., Loménie, N.: Learning an adaptation function to assess image visual similarities. In: 2021 IEEE International Conference on Image Processing (ICIP), pp. 2498–2502 (2021). https://doi.org/10.1109/ICIP42928.2021.9506129
19. Rosenfeld, A., Solbach, M.D., Tsotsos, J.K.: Totally looks like - how humans compare, compared to machines. In: 2018 IEEE/CVF Conference on Computer Vision and Pattern Recognition Workshops (CVPRW), pp. 2042–20423. IEEE Computer Society, June 2018
20. Roy, S., Harandi, M., Nock, R., Hartley, R.: Siamese networks: the tale of two manifolds. In: 2019 IEEE/CVF International Conference on Computer Vision (ICCV), pp. 3046–3055 (2019). https://doi.org/10.1109/ICCV.2019.00314
21. Sanders, C., Nosofsky, R.: Training deep networks to construct a psychological feature space for a natural-object category domain. Comput. Brain Behav. **03**, 229–251 (2020). https://doi.org/10.1007/s42113-020-00073-z
22. Santoro, A., Bartunov, S., Botvinick, M.M., Wierstra, D., Lillicrap, T.P.: One-shot learning with memory-augmented neural networks. CoRR abs/1605.06065 (2016). arxiv:1605.06065
23. Taylor, J., Xu, Y.: Representation of color, form, and their conjunction across the human ventral visual pathway. NeuroImage **251**, 118941 (2022). https://doi.org/10.1016/j.neuroimage.2022.118941
24. Zhang, K., Zhang, Z., Li, Z., Qiao, Y.: Joint face detection and alignment using multitask cascaded convolutional networks. IEEE Signal Process. Lett. **23**(10), 1499–1503 (2016). https://doi.org/10.1109/LSP.2016.2603342
25. Zhang, R., Isola, P., Efros, A.A., Shechtman, E., Wang, O.: The unreasonable effectiveness of deep features as a perceptual metric. In: 2018 IEEE/CVF Conference on Computer Vision and Pattern Recognition (CVPR), pp. 586–595, June 2018

Applications

Wavelet Scattering Transform Depth Benefit, An Application for Speaker Identification

Abderrazzaq Moufidi[1,2], David Rousseau[1(✉)], and Pejman Rasti[1,2]

[1] Université d'Angers, LARIS, UMR INRAe IRHS, Angers, France
abderrazzaq.moufidi@etud.univ-angers.fr,
{david.rousseau,pejman.rasti}@univ-angers.fr
[2] CERADE, ESAIP, Angers, France

Abstract. This paper assesses the interest of the multiscale Wavelet Scattering Transform (WST) for Speaker identification (SI) applied in several depths and invariance scales. Our primary purpose is to present an approach to optimally design the WST to enhance the identification accuracy for short utterances. We describe the invariant features offered by the depth of this transform by performing simple experiments based on text-independent and text-dependent SI. To compete the state-of-the-art (SOTA), we propose a fusion method between WST and x-vectors architecture, we show that this structure outperforms HWSTCNN by 7.57% on TIMIT dataset sampled at 8 kHz and makes the same performance in the SOTA at 16 kHz.

Keywords: Wavelet scattering transform · x-vectors · Speaker identification · Invariance scale · Depth

1 Introduction

Wavelet Scattering Transform (WST) is a scattering cascade of convolutions and non-linearities based on wavelets and proposed by S. Mallat [1,8]. It was inspired by LeNet architecture [5] and the wavelet transform [7], it performs an extraction of features called scatter coefficients at different scales by using wavelets instead of learnable filters used in standard CNN, which makes it more understandable and explainable than CNNs [9]. The WST is locally translation invariant and stable to random deformations up to a log term [8], thus it is more robust to some extrinsic variations in speaker identification tasks such as reverberation.

A broad range of patterns recognition problems have been studied with WST since its introduction in 2011 including iris recognition [10], rainfall classification in radar images [4], cell-scale characterization [6], weed detection [12] or speaker identification [3]. Moreover, the WST has proven to be effective in these applications; however, its coefficients were not systematically explored in voice tasks for more than the second-order before feeding the features to a classifier, specifically where voice signals considered as the input to the model.

© The Author(s), under exclusive license to Springer Nature Switzerland AG 2023
N. El Gayar et al. (Eds.): ANNPR 2022, LNAI 13739, pp. 97–106, 2023.
https://doi.org/10.1007/978-3-031-20650-4_8

In this paper, we explore the importance of the depth and the invariance scale of WST [1] in an application on the speaker identification task on TIMIT dataset [2], for both text dependent and independent tasks under the conditions, shortness of the utterances and the small value of the sampling frequency.

2 Related Work

The concept of WST method is to apply iteratively the wavelet transform ψ and modulus as a non-linearity function and an average Gaussian filter ϕ. An audio signal x is convoluted $\star$ with dilated wavelets ψ_λ generated from a mother wavelet:

$$\psi_{\lambda_i}(t) = \lambda_i \psi(\lambda_i t), \tag{1}$$

where $\lambda_i = 2^{\frac{j}{Q_i}}$, $j \in \mathbb{Z}$ and Q_i is the quality factor or the number of wavelets per octave and $i > 0$ refers to the order index. ψ_{λ_i} are centered at λ_i and have a bandwidth of $\frac{\lambda_i}{Q_i}$ in the frequency domain.

At the zero-order, the scatter coefficients are given by $S_0(x) = x \star \phi(t)$. At the i−th order ($i \geq 1$), the signal is down-sampled by using the mean pooling at every scale $i \geq 2$, and the scatter coefficients are calculated as follows until a fixed maximum order m

$$S_i x(t, \lambda_1, ..., \lambda_i) = |||x \star \psi_{\lambda_1}| \star ...| \star \psi_{\lambda_i}| \star \phi(t). \tag{2}$$

These coefficients $(S_i x)_{i=1,..,m}$ are often log-normalised to reduce redundancy and increase translation invariance as given in the following equations:

$$\widetilde{S}_1 x (t, \lambda_1) = \log\left(\frac{S_1 x (t, \lambda_1)}{|x| \star \phi(t) + \epsilon}\right) \tag{3}$$

$$\widetilde{S}_i x(t, \lambda_1, ..., \lambda_i) = \log\left(\frac{S_i x(t, \lambda_1, ..., \lambda_i)}{S_{i-1} x(t, \lambda_1, ..., \lambda_{i-1}) + \epsilon}\right) \tag{4}$$

where ϵ is a silence detection threshold and $i \geq 2$ is the depth that we search to optimize with another parameter called the invariance scale i.e. the frame length of the Gaussian filter ϕ presented in Eq. 2.

The wavelets at the first order are mel-scaled to cover human frequencies by taking $Q_1 = 8$ (more selective at low frequencies and less at high frequencies), and the quality factors for depths $i \geq 2$ are set to $Q_i = 1$. In audio tasks [1,3], the invariance scale was always set to 32 ms, and the classification was limited to the second depth order based on the fact that the most amount of energy is absorbed by the first and second-order scatter coefficients (cf. Tables 1 and 2). Despite that at this depth and invariance, WST surpasses MFCCs in phone or music recognition task [1], there were no investigations on the depth benefit in speaker recognition. In a computer vision task, [12] found that the power of WST results in its deepness. It was proved that the energy amount absorbed by orders does not give discriminant information between the classes. Instead, it is better to look for the optimum WST depth that maximizes the contrast between two different classes.

Table 1. Energy absorbed at different scale for some invariance scale T and N_s the total number of scatter coefficients for TIMIT audios sampled at 8 kHz.

$\frac{||Sx||^2}{||x||^2}\%$	Depth				
$T\|\|N_s$	$1st$	$2nd$	$3rd$	$4th$	$5th$
$16\,ms\|\|35$	90.18	0.21	–	–	–
$32\,ms\|\|74$	86.69	0.61	$4.5e-3$	–	–
$64\,ms\|\|152$	82.79	1.21	$2.08e-2$	$2.13e-4$	–
$128\,ms\|\|308$	71.04	3.39	$6.1e-2$	$1.4e-3$	$1.38e-5$

Table 2. Energy absorbed at different scale for some invariance scales T and N_s the total number of scatter coefficients for TIMIT audios sampled at 16 kHz.

$\frac{||Sx||^2}{||x||^2}\%$	Depth					
$T\|\|N_s$	$1st$	$2nd$	$3rd$	$4th$	$5th$	$6th$
$16\,ms\|\|74$	89.65	0.54	$4.0e-3$	–	–	–
$32\,ms\|\|152$	86.13	1.01	$2.17e-2$	$2.13e-4$	–	–
$64\,ms\|\|308$	82.18	1.6	$5.05e-2$	$1.4e-3$	$1.32e-5$	–
$128\,ms\|\|620$	70.47	3.72	0.11	$4.4e-3$	$1.11e-4$	$9.34e-7$

To the best of our knowledge, no investigations were conducted on the benefits of the depths and the invariance scale of the WST in the speaker identification for both text-independent and text-dependent task. To measure the impact of these two factors on the classification accuracy per sentence and the reduction of learnable parameters for text-independent identification on TIMIT dataset [2]. Three architectures will serve as baselines: CNN-raw sample [11], SincNet [13] and HWSTCNN [3]. Also, we will show visually the benefits of depths > 2 when performing the speaker text-dependent identification.

2.1   CNN-raw System

CNN-raw system is an end-to-end speaker recognition method [11], in other words a raw sample is fed directly to a CNN, the first layer performs a set of time domain convolutions between 200 ms chunks with an overlapping of 10 ms and 80 standard CNN filters of kernel size 251, followed by filters of kernel size 5 and two fully connected layers of 2048.

2.2   SincNet

SincNet is an end-to-end novel network proposed by Mirco Ravanelli [13], it is similar to the CNN-raw architecture except at the first layer where it performs a set of time domain convolutions between 200 ms chunks with an overlapping of 10 ms and 80 Shannon wavelet filters of length 251, these filters were initialized by mel-scaled frequencies, therefore the parameters are considerably reduced compare to the original CNN raw architecture presented in [11].

2.3   HWSTCNN

Hybrid Wavelet Scattering Transform Convolutional Neural Network (HWSTCNN) [3] is an end-to-end speaker identification method composed of a WST applied with 32 ms invariance scale until the second depth, fused with a CNN, the architecture receives an audio decomposed to 500 ms chunks with 250 ms overlapping.

3   Experimental Setup

To find the optimal values of depth and invariance scale of WST that are suited for speaker identification (SI), we perform SI text-independent on 462 speakers from TIMIT dataset and SI text-dependent on two speakers reading the same sentence. The method is fused with a CNN and compared to the three baselines.

3.1   Speaker Identification Text-Independent

Depth and Invariance Scale. In order to extract the optimum depth and invariance scale of WST for SI task, we work on SI text-independent task done on 462 speakers from TIMIT [2] sampled at 8 kHz and 16 kHz. Firstly, we preprocess the audio files by removing the silent frames at the beginning and the end of an utterance (no pre-emphasize was applied). Secondly, we apply the WST by using different invariance scales 16,32,64,128 ms until their maximum depth. We omitted the WST of silent frames for this experiment. The WST coefficients are log-normalized following Eq. 4. The resulted frames from 5 utterances beginning with *"sx"* are used for the training phase and the last 3 utterances starting with *"si"* are used for the testing part. A multilayer perceptron (MLP) is used as a classifier with cross entropy as a loss function. The batch size and the maximum epoch were set respectively to 256 and 100. The optimizer used was ADAMAX with a learning rate 2.10^{-2}.

Comparison to Baselines. To enhance the classification accuracy per sentence, feeding WST coefficients to a neural network, inspired from x-vectors architecture [14], is essential to have performance as the SOTA. At the input, the architecture given in Fig. 1 receives WST coefficients of 230 ms frame length and an overlapping of 58 ms. At the testing phase, we average the probabilities resulted from each frames of a given sentence to give the corresponding speaker. The experiment was conducted for different depths for each invariance scale 16, 32, 64 ms, in order to observe the effect of WST depth under CNN fusion. The architecture was trained using cross-entropy loss, the batch size was set to 256, the maximum epoch was set to 100. The optimizer used was ADAMAX with a learning rate 5.10^{-3}.

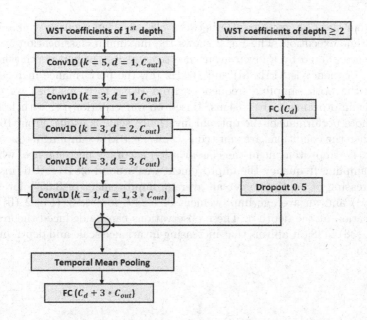

Fig. 1. Fusion of x-vectors architecture with WST (C_{out} is the number of output channels from a 1D temporal convolution, here it was set to 128, C_d is the number of WST coefficients of depth ≥ 2).

3.2 Speaker Identification Text-Dependent

We show visually the benefit of depths > 2 in speaker identification text-dependent task. We choose two speakers from TIMIT dataset reading the sentence *"she had your dark suit in greasy wash water all year"* under a clean speaking record environment, then we apply the WST for the optimal depth and invariance scales values found previously. To observe the impact of depth, we plot the spectrum at each depth for each speaker. One can say; the magnitude of the WST coefficients depends on the intensity of the speaker when he pronounces a word, therefore for visualization reasons, we choose to normalize (max-normalization) our data per scale in order to see which interference value is maximum for a speaker. The color-bar scale was adjusted to clearly visualize the differences. The WST coefficients were not log-normalized as in the previous part.

4   Results and Discussion

4.1   Speaker Identification Text-Independent

Optimal Invariance Scale and Depth. To highlight the impact of the WST invariance scale and its depth on TIMIT dataset, we used a different combination of invariance scales (ms) and the number of depths to realize which structure best fits our data i.e. compromise between classification accuracy per sentence and the time execution. The Fig. 2 shows the maximum classification accuracy per sentence offered by a given invariance scale at its optimal depth when the sampling frequency is 8 kHz (a) and 16 kHz (c), the performance increases linearly for the both sampling frequencies from 16 ms to 64 ms then we do not see bigger improvement after 64 ms. Based on our criterion i.e. with less time we get more performance, the optimal invariance scale is 64 ms. From 16 ms to 64 ms, when the conditions are more constraint, i.e. 8 kHz sampling frequency, we see a 18.04% improvement in the classification accuracy per sentence, while for 16 kHz sampling frequency the improvement was around 8.87%. At 64 ms, after each increasing of depth, we get an average improvement of 3.05% for 16 kHz (Fig. 2 (a)) and an average improvement of 5.24% for 8 kHz (Fig. 2 (b)) until a stabilisation at the depth 4. These observations can be deduced theoretically from Eq. (38) in [8], it affirms that increasing invariance scale and depth increase invariance.

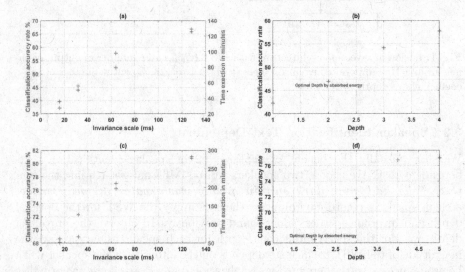

Fig. 2. Classification accuracy per sentence for different invariance scales when the audios are sampled at 16 kHz (c) and at 8 kHz (a), and for different depths of the best invariance scale 64 ms sampled at 16 kHz (d) and at 8 kHz (b).

WST+x-vectors. From the results shown in Fig. 2, a fusion of WST with a CNN is crucial to increase the classification accuracy. CNNs captures details from the first depth of WST and depths ≥ 2 gives an information on large structures. In the Fig. 3 we report the results of the WST+x-vectors (Fig. 1) applied for different depths and invariance scales. The optimal invariance scale for WST+x-vectors that increases the classification accuracy per sentence is 16 ms instead of 64 ms, this can be explained by the locality of CNN and the temporal high resolution offered by 16 ms. Depending on the invariance scale used the optimal

WST depth that improves the classification accuracy per sentence differs, yet increasing it always leads to an improvement of the classification accuracy per frame. To evaluate our architecture, a performance overview and the number of learnable parameters of the baselines and the WST+x-vectors are given in the table 3. Compared to the baselines methods, the number of learnable parameters required by our architecture is less by 94%, yet it outperforms HWSTCNN under 8 kHz sampling frequency condition by an improvement of 7.57%, and makes the same order of performance compared to the baselines under 16 kHz sampling frequency.

Table 3. Classification accuracy per sentence of the best values of invariance scale and depth of WST+x-vectors and baselines trained and tested on TIMIT sampled at 8 kHz and 16 kHz (#*param* is the number of learnable parameters).

Architecture	8 kHz	16 kHz
CNN-raw (#*param* = 22.8M)	**97.75%**	98.91%
SincNet (#*param* = 22.7M)	97.54%	**99.4%**
HWSTCNN (#*param* = 18.1M)	85.93%	98.12%
WST + x-vectors (#*param* < 1M)	93.5%	98.12%

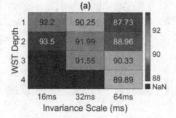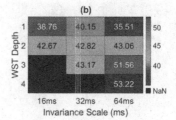

Fig. 3. The classification accuracy per sentence (a) and the classification accuracy per frame (b) for TIMIT sampled at 8 kHz when using WST + x-vectors for different depths of invariance scales ≤ the optimal invariance scale 64 ms.

4.2   Speaker Identification Text-Dependent

To understand visually the behavior of WST across depths, we performed a comparison between the WST coefficients of two speakers reading the same sentence. Figure 4 is the spectrogram of the scatter coefficients of the first and the second orders of the two speakers, the first depth presents two main peaks approximatively located at $200Hz$ which represents the pitch contour, and the second one at around $425Hz$. At the second depth, the invariant features are more enhanced compared to the first depth. At depths > 2 in Fig. 5 for a given word, the distribution along frequencies differs strongly from a speaker to another, therefore going deeper generates more invariant features that their distribution along frequencies depends on speaker's identity.

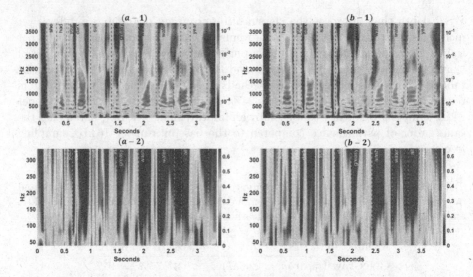

Fig. 4. Spectrogram of the WST coefficients at the first depth (first speaker (a-1) and the second speaker (b-1)) and the second depth (first speaker (a-2) and the second speaker (b-2)) (x-axis time and y-axis is frequencies).

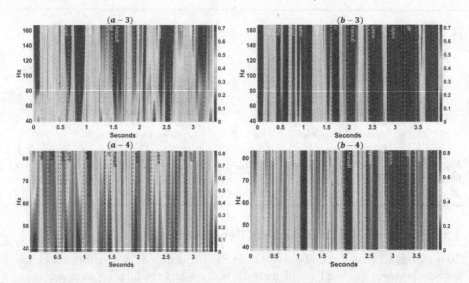

Fig. 5. Spectrogram of the WST coefficients at the third depth (first speaker (a-3) and the second speaker (b-3)) and the fourth depth (first speaker (a-4) and the second speaker (b-4)) (x-axis time and y-axis is frequencies).

5   Conclusion

In this paper, we have shown the importance of WST depth and its invariance scale for speaker identification. We have proved that instead of looking for energy concentration at the first two depths of WST, we should go deeper to generate invariant features. Experimental results on TIMIT have shown that the deeper WST can achieve dominant results with limited data. An optimized method based on a compromise between the classification accuracy per sentence and execution time has been successfully proposed to select a priori the best scatter transform architecture for speaker identification text-independent task. To enhance the classification accuracy per sentence and compete the SOTA, we have proposed a fusion between CNN and WST and we have proven that the WST depth increases the classification accuracy per frame. The resulted optimal values of WST invariance scale and depth, were used to observe visually the benefits in a speaker identification text-dependent task. These results show significant promise for considerable improvement in speaker identification. As a possible perspective of the investigation, one could further optimize the quality factors (the number of wavelet filters per octave) for high-scale orders of WST. In most experiments, the quality factors of the first two layers are fixed (8 and 1 respectively for audio and speech signal processing), but there is no investigation for other layers of WST and their impact on classification problems.

References

1. Andén, J., Mallat, S.: Deep scattering spectrum. IEEE Trans. Signal Process. **62**(16), 4114–4128 (2014). https://doi.org/10.1109/TSP.2014.2326991
2. Garofolo, J.S.: Timit acoustic phonetic continuous speech corpus. Linguistic Data Consortium, 1993 (1993)
3. Ghezaiel, W., Brun, L., Lézoray, O.: Hybrid network for end-to-end text-independent speaker identification. In: 2020 25th International Conference on Pattern Recognition (ICPR), pp. 2352–2359 (2021). https://doi.org/10.1109/ICPR48806.2021.9413293
4. Lagrange, M., Andrieu, H., Emmanuel, I., Busquets, G., Loubrié, S.: Classification of rainfall radar images using the scattering transform. J. Hydrol. **556**, 972–979 (2018)
5. LeCun, Y., et al.: Backpropagation applied to handwritten zip code recognition. Neural Comput. **1**(4), 541–551 (1989)
6. Li, B.H., Zhang, J., Zheng, W.S.: Hep-2 cells staining patterns classification via wavelet scattering network and random forest. In: 2015 3rd IAPR Asian Conference on Pattern Recognition (ACPR), pp. 406–410. IEEE (2015)
7. Mallat, S.: A wavelet tour of signal processing. Elsevier (1999)
8. Mallat, S.: Group invariant scattering. Commun. Pure Appl. Math. **65**(10), 1331–1398 (2012)
9. Mallat, S.: Understanding deep convolutional networks. Philos. Trans. R. Soc. A: Math. Phys. Eng. Sci. **374**(2065), 20150203 (2016)
10. Minaee, S., Abdolrashidi, A., Wang, Y.: Iris recognition using scattering transform and textural features. In: 2015 IEEE signal processing and signal processing education workshop (SP/SPE), pp. 37–42. IEEE (2015)

11. Muckenhirn, H., Magimai-Doss, M., Marcel, S.: On learning vocal tract system related speaker discriminative information from raw signal using CNNs. In: INTERSPEECH, pp. 1116–1120 (2018)
12. Rasti, P., Ahmad, A., Samiei, S., Belin, E., Rousseau, D.: Supervised image classification by scattering transform with application to weed detection in culture crops of high density. Remote Sens. **11**(3), 249 (2019). https://doi.org/10.3390/rs11030249. https://www.mdpi.com/2072-4292/11/3/249
13. Ravanelli, M., Bengio, Y.: Speaker recognition from raw waveform with sincNet. In: 2018 IEEE Spoken Language Technology Workshop (SLT), pp. 1021–1028 (2018). https://doi.org/10.1109/SLT.2018.8639585
14. Snyder, D., Garcia-Romero, D., Sell, G., Povey, D., Khudanpur, S.: X-vectors: Robust DNN embeddings for speaker recognition. In: 2018 IEEE International Conference on Acoustics, Speech and Signal Processing (ICASSP), pp. 5329–5333 (2018). https://doi.org/10.1109/ICASSP.2018.8461375

Sequence-to-Sequence CNN-BiLSTM Based Glottal Closure Instant Detection from Raw Speech

Jindřich Matoušek[1,2]([⊠])[iD] and Daniel Tihelka[2][iD]

[1] Department of Cybernetics, Faculty of Applied Sciences, University of West Bohemia, Plzeň, Czech Republic
jmatouse@kky.zcu.cz
[2] New Technology for the Information Society (NTIS), Faculty of Applied Sciences, University of West Bohemia, Plzeň, Czech Republic
dtihelka@ntis.zcu.cz

Abstract. In this paper, we propose to frame glottal closure instant (GCI) detection from raw speech as a sequence-to-sequence prediction problem and to explore the potential of recurrent neural networks (RNNs) to handle this problem. We compare the RNN architecture to widely used convolutional neural networks (CNNs) and to some other machine learning-based and traditional non-learning algorithms on several publicly available databases. We show that the RNN architecture improves GCI detection. The best results were achieved for a joint CNN-BiLSTM model in which RNN is composed of bidirectional long short-term memory (BiLSTM) units and CNN layers are used to extract relevant features.

Keywords: Glottal closure instant detection · Deep learning · Recurrent neural network · Convolutional neural network

1 Introduction

Pattern recognition and especially *deep learning* using *artificial neural networks* (ANNs) has recently been successfully applied in many areas of signal processing, replacing the established and refined signal processing techniques (such as autocorrelation, convolution, Fourier and wavelet transforms and many others), or speech/audio modeling and processing techniques (such as Gaussian mixture models or hidden Markov models) [26]. It is also the case of *glottal closure instant*

This work was supported by the Technology Agency of the Czech Republic (TA CR), project No. TL05000546, and from ERDF "Research and Development of Intelligent Components of Advanced Technologies for the Pilsen Metropolitan Area (InteCom)" (No. CZ.02.1.01/0.0/0.0/17_048/0007267). Computational resources were supplied by the project "e-Infrastruktura CZ" (e-INFRA CZ LM2018140) supported by the Ministry of Education, Youth and Sports of the Czech Republic.

N. El Gayar et al. (Eds.): ANNPR 2022, LNAI 13739, pp. 107–120, 2023.
https://doi.org/10.1007/978-3-031-20650-4_9

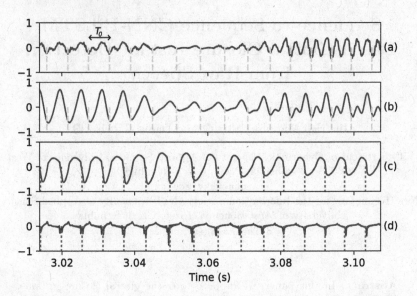

Fig. 1. An example of a raw speech signal (a), its low-pass filtered representation (b) used in the baseline system described in Sect. 3.3, EGG signal (c), and its difference (d). GCIs (corresponding to negative peaks) are marked by red dashed and green dotted lines in speech-based and EGG-based signals respectively (note the slight delay between speech and EGG signals caused by the different places in the vocal tract where the signals are measured: larynx or lips).). T_0 denotes the local pitch period. (Color figure online)

detection, a traditional signal processing/detection task. Deep *convolutional neural networks* (CNNs) have been shown to beat traditionally used algorithms for glottal closure instant detection [11] (such as SEDREAMS [10] or DYPSA [24]) [4,12,16,23,27].

Detection of glottal closure instants (GCIs) could be viewed as a task of pattern recognition. It aims at determining peaks in the *voiced parts* of the speech signal that correspond to the moment of glottal closure, a significant excitation of the vocal tract during speaking. The distance between two succeeding GCIs then corresponds to one vocal fold vibration cycle and can be represented in the time domain by a local *pitch period* value (T_0) or in the frequency domain by a local *fundamental frequency* value ($F_0 = 1/T_0$). The sequence of GCIs describes a pitch of a given utterance. Note that GCIs are present only in voiced segments of speech as there is no vocal fold vibration in unvoiced speech segments. An example of the voiced speech signal and the corresponding GCIs is shown in Fig. 1a,b.

Accurate location of GCIs can be beneficial in many practical applications, especially in those where *pitch-synchronous* speech processing is required such as in pitch tracking, data-driven voice source modeling, speech enhancement and dereverberation, glottal flow estimation and speaker recognition, prosodic speech modification, or in various areas of speech synthesis and voice conversion/transformation [9, 11, 33, 34].

Although GCIs can be reliably detected from a simultaneously recorded electroglottograph (EGG) signal (which measures glottal activity directly; thus, it is not burdened by modifications that happen to a flow of speech in the vocal tract – see Fig. 1c, d), it is not always possible (e.g. in the case of using existing speech recordings) or comfortable to use an EGG device during the recording. Hence, there is a great interest to detect GCIs directly from the raw speech signal.

Traditional algorithms usually exploit expert knowledge and hand-crafted rules and thresholds to identify GCI candidates from local maxima of various speech representations and/or from discontinuities or changes in signal energy. The former include linear predictive coding (e.g. DYPSA [24], YAGA [34], or [25]), wavelet components [35], or multiscale formalism (MMF) [15]. The latter include Hilbert envelope, Frobenius norm, zero-frequency resonator, or SEDREAMS [10]. Dynamic programming was often used to refine the GCI candidates [24, 34]. A universal postprocessing scheme to correct GCI detection errors was also proposed [32]. A nice overview of the algorithms can be found in [11].

More recently, a machine-learning-based approach prevailed [21]. From the point of view of pattern recognition, GCI detection could be described as a two-class classification problem: whether or not a peak in a speech waveform represents a GCI [5]. The advantage of a machine-learning-based approach is that once a training dataset is available and relevant features are identified from raw speech, classifier parameters are set up automatically without manual tuning. On the other hand, the identification of relevant features may be time-consuming and tricky, especially when carried out by hand [22]. Unlike classical ("non-deep") machine learning, deep learning, and especially CNNs, can help solve the problem of identifying features. In general, deep learning can help in finding more complex dependencies between raw speech and the corresponding GCIs. CNNs can directly be applied to the raw speech signal without requiring any pre- or post-processing, such as feature identification, extraction, selection, dimension reduction, etc. – steps that must be carried out in the case of classical machine learning [7, 16, 21]. CNN-based architectures were very popular in previous GCI detection studies [4, 12, 16, 23, 27]; however, only a few studies seem to have investigated recurrent architectures in the context of deep learning-based GCI detection [31].

In this paper, we propose to frame GCI detection as a sequence-to-sequence prediction problem and we explore the potential of *recurrent neural networks* (RNNs), specifically their gated variants, *long short-term memory* (LSTM) networks and *gated recurrent units* (GRUs) to handle this problem. We also examine a joint CNN-BiLSTM architecture which involves using CNN layers for feature extraction on raw speech data combined with *bidirectional* LSTMs to support sequence prediction.

2   Data Description

2.1   Speech Material

Experiments were performed on clean 16 kHz sampled speech recordings primarily intended for speech synthesis. We used 3200 utterances from 16 voice talents (8 male and 8 female voices with 200 utterances per voice) of different languages (8 Czech, 2 Slovak, 3 US English, Russian, German, and French). Two voices were from the CMU ARCTIC database [2,17] (Canadian English JMK and Indian English KSP), the remaining voices were borrowed from our partner company. For our purposes, speech waveforms were mastered to have equal loudness and negative polarity (dominant peaks are under zero) [20]. 3136 utterances (196 from each voice) were used for training, the rest was used for tuning and validation.

Since it is very hard and time-consuming to mark all 3200 speech signals by hand, we decided to use GCIs detected by the Multi-Phase Algorithm (MPA) [19] from contemporaneous EGG recordings as ground truths. They may not be 100% correct but they are very accurate as the detection from less complex EGG signals is a much easier task (see e.g. [19]). Due to the slight delay between EGG and speech signals, the EGG-detected GCIs were shifted towards the neighboring minimum negative sample in the speech signal.

2.2   GCI Detection Measures

GCI detection techniques are usually evaluated by comparing the locations of the detected and reference GCIs. The measures typically concern the *reliability* and *accuracy* of the GCI detection algorithms [24]. The former includes the percentage of glottal closures for which exactly one GCI is detected (*identification rate*, IDR), the percentage of glottal closures for which no GCI is detected (*miss rate*, MR), and the percentage of glottal closures for which more than one GCI is detected (*false alarm rate*, FAR). The latter includes the percentage of detections with the identification error $\zeta \leq 0.25$ ms (*accuracy to* ± 0.25 *ms*, A25) and standard deviation of the identification error ζ (*identification accuracy*, IDA).

In addition, we use a more *dynamic evaluation measure* [18]

$$E10 = \frac{N_{GT} - N_{\zeta > 0.1T_0} - N_M - N_{FA}}{N_{GT}} \tag{1}$$

that combines the reliability and accuracy in a single score and reflects the local *pitch period* T_0 pattern (determined from the ground truth GCIs). N_{GT} stands for the number of reference GCIs, N_M is the number of missing GCIs (corresponding to MR), N_{FA} is the number of false GCIs (corresponding to FAR), and $N_{\zeta > 0.1T_0}$ is the number of GCIs with the identification error ζ greater than 10% of the local pitch period T_0. For the alignment between the detected and ground truth GCIs, dynamic programming was employed [18].

As we consider reliability more important than accuracy (we prefer better identification over absolute accuracy in GCI location), the proposed models were tuned with respect to IDR (and E10) measures.

3 Models

3.1 Baseline CNN-Based GCI Detection System

We used the CNN-based GCI detection architecture proposed in [23] as the baseline system. Specifically, we used a one-dimensional InceptionV3-1D model that achieved the best GCI detection results.

Since CNNs predict each GCI independently on previous/next GCIs, the detection of peaks as GCI/non-GCI can be carried out in a *peak-by-peak manner* [22,23]. In this scenario, negative peaks were detected by zero-crossing low-pass filtered (by a zero-phase Equiripple-designed filter with 0.5 dB ripple in the pass band, 60 dB attenuation in the stop band, and the cutoff frequency 800 Hz) speech signal exactly in the same way as described in [22] (for an example of the low-pass filtered signal see Fig. 1b or the red dashed line in Fig. 2). It was also found that downsampling to 8 kHz prior to filtering provided slightly better results than the use of 16 kHz. Thus, the baseline InceptionV3-1D GCI detection model use 8 kHz internally. The block diagram of the peak-by-peak CNN GCI detection is shown in Fig. 2. We also tried to employ a *frame-based* detection (explained further in Sect. 3.2) but we got worse results.

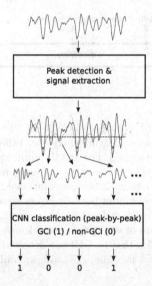

Fig. 2. A simplified scheme of the peak-by-peak CNN GCI detection from raw speech signal (blue solid line). Negative peaks (either true GCIs ● or non-GCIs ○) are detected from the corresponding low-pass filtered signal (red dashed line). (Color figure online)

3.2 Recurrent Neural Network-Based GCI Detection

While convolutional neural networks were shown to perform well on the GCI detection task, their disadvantage is that they do not take the temporal depen-

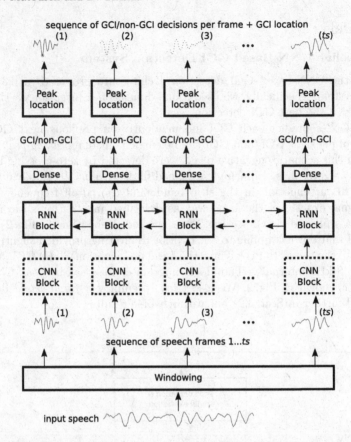

Fig. 3. A simplified scheme of the proposed CNN-BiLSTM-based GCI detection. The CNN block (dotted line) works as a feature extractor. When omitted, RNN-based GCI detection on raw speech is performed. For 16 kHz input speech, three BiLSTM layers with 256 cells in each layer and 900 time steps (ts) were used in RNN blocks. In CNN blocks, three convolutional blocks with two convolutional layers in each block followed by batch normalization and maximum pooling layers were used (with the number of filters 16, 32, 64, kernel size 7 with stride 1, pooling size 3, and "same" padding). The dense layer outputs a prediction of whether or not a frame contains a GCI. The dotted speech signals at the top indicate that no GCI was detected in the corresponding speech frames; otherwise, • marks GCI location. (Color figure online)

dencies of GCIs (and the temporal structure of speech in general) into account. Since vocal folds vibrate during speaking in a quasiperiodic way, generating a GCI on each glottal closure, there is a temporal pattern present in the resulting speech signal which is not captured by the CNN architecture.

On the other hand, recurrent neural networks (RNNs) are capable of capturing the temporal structure present in the input time series data [28,30]. The input speech signal can be viewed as a sequence of frames consisting of speech samples, such that each frame depends on previous (in the case of a bidirectional architecture also on next) frames, and RNNs can then incorporate the dependencies between these speech frames. RNNs are often thought to have the concept of "memory" (or internal state) that helps them to store the states or information of previous (and next in the case of the bidirectional architecture) inputs to generate the actual output of the sequence.

Unlike the CNN-based detection, where GCIs were detected independently of each other in a *peak-by-peak* manner (see Sect. 3.1), a *frame-based* detection was carried out for the RNN-based detection to capture the temporal structure of the input speech signal. In this way, the speech signal was divided into overlapping frames using a sliding window of a given length w and given hop length h. Note that no speech filtering and peak detection is performed here. Frame-based detection proved to be better than the peak-based one (see Table 1).

Table 1. Comparison of RNN-based GCI detection on the validation set, including the approximate number of models' parameters. The model name consists of an RNN unit, number of layers, number of cells in each layer, sampling frequency in kHz, and frame- or peak-based detection.

Model	IDR (%)	A25 (%)	E10 (%)	# params
BiLSTM3-256–16f	**97.24**	97.76	**95.89**	4.0 M
BiLSTM3-64–8f	97.16	97.58	95.86	0.3 M
BiGRU3-128–16p	96.01	98.60	94.92	0.8 M
BiLSTM3-256-8p	95.89	**98.62**	94.82	4.0 M

So, using RNNs, GCI detection could be viewed as a *sequence-to-sequence prediction problem*. In this framework, for each input sequence of speech frames, an output sequence of the same length is predicted assigning to each frame a prediction of whether or not the frame contains a GCI. If the frame contains the GCI, the minimum negative sample in the frame is selected as the GCI. The length of the sequences is often referred to as a number of *time steps*. If the input signal contained fewer frames than the given number of time steps, it was zero-padded accordingly. As it is well-known that simple RNNs are prone to training problems known as vanishing or exploding gradient, we used their gated variants, *long short-term memory* (LSTM) networks [13] and *gated recurrent units* (GRU) [6], to alleviate the problems. Finally, a *dense* (fully connected) layer is stacked on top of the recurrent layers to output a prediction. A simplified scheme of RNN-based GCI detection is given in Fig. 3.

There are several hyper-parameters that should be experimented with when training an RNN model for our purposes. The following ones were taken into account in our comparison and tuned on the validation set: sampling frequency ($sf = \{8, 16\}$ kHz), window type ($t = \{$rectangular, von Hann$\}$), window length ($w = 2$–24 ms), number of time steps ($ts = 10$–1400), RNN type ($r = \{$LSTM, GRU$\}$), number of RNN cells ($c = 16$–1280), number of recurrent layers ($l = 1$–3), learning rate ($lr = 0.000001$–0.01), mini-batch size ($bs = 1$–128), and dropout

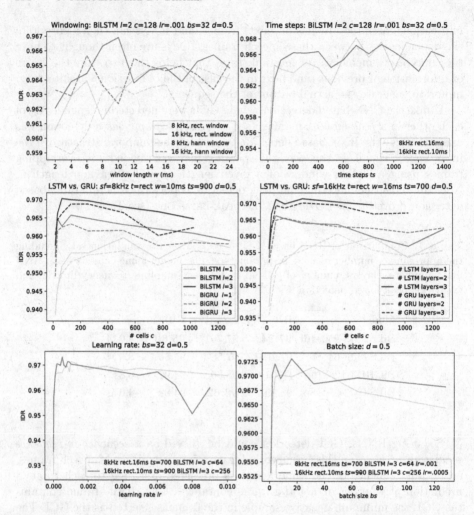

Fig. 4. The influence of different hyper-parameters on the RNN-based GCI detection performance (in terms of IDR).

to avoid overfitting ($d = 0.0$–0.9 with $d = 0.5$ giving the best results). The hop length was set to $h = 2$ ms as this value corresponds to the minimum possible pitch period (assuming that the highest vocal fold frequency in our data 500 Hz). The influence of different hyper-parameters on the GCI detection performance is shown in Fig. 4. Briefly, lower learning rates and mini-batch sizes and a higher number of time steps are preferred, 3-layer architectures with LSTMs are better, smaller (for 16 kHz) and longer (for 8 kHz) rectangular windows are a good choice.

In all experiments, the networks were trained to minimize a *binary cross-entropy loss* using *mini-batch gradient descent* with the *Adam optimizer*. Default activation functions, *tanh* and *sigmoid*, were applied in the recurrent layers, and *sigmoid activation* was used in the last (dense) layer. To speed up the training, it was stopped when the validation loss did not improve for 10 epochs and the maximum number of epochs was set to 100. Bidirectional versions of the recurrent models, i.e. BiLSTM and BiGRU, were used.

3.3 CNN-BiLSTM GCI Detection

In the next series of experiments, we examined a joint CNN-RNN architecture in which the feature extraction power of CNNs is combined with the ability of RNNs to capture the temporal structure of the input time series data and to model temporal dependencies between a sequence of speech frames [8]. Specifically, CNNs were used to extract GCI detection-relevant features from input raw speech data, and simultaneously, RNNs were used both to interpret the features across time steps and to detect GCIs.

We experimented with several CNN architectures CNNn where n is a number of convolutional blocks. Each block typically consisted of two convolutional layers followed by batch normalization, dropout, and maximum pooling layers. We also tried some more complex models – the InceptionV3-1D model which yielded the best results in the CNN-based GCI detection [23], see Sect. 3.1, 1D version of VGG11 (a lightweight version of the well-known VGG architecture proposed for image processing [29]), and SwishNet which was proposed directly for audio processing [14].

Table 2. Comparison of CNN-RNN GCI detection on the validation set with different CNN models and the best RNN model (BiLSTM3-256–16f for 16 kHz and BiLSTM3-64–8f for 8 kHz), including the approximate number of models' parameters. The last number in model names denotes sampling frequency in kHz.

Model	IDR (%)	A25 (%)	E10 (%)	# params
CNN3-16	**97.49**	98.95	**96.68**	4.4 M
CNN3-8	97.37	99.03	96.63	4.5 M
InceptionV3-16	97.41	99.01	96.65	23.9 M
InceptionV3-8	96.45	99.04	95.73	21.3 M
SwishNet-16	96.96	99.00	96.20	16.8 M
SwishNet-8	96.54	99.02	95.81	14.2 M
VGG11-16	96.14	99.01	95.35	12.0 M
VGG11-8	95.68	**99.07**	94.95	10.9 M

4 Results

4.1 Comparison of Proposed Models

As can be seen in Table 1 and Fig. 4, the best results for the RNN-based detection described in Sect. 3.2 were achieved for the 3-layered BiLSTMs with 256 cells in each layer and 16 kHz frame-based speech input (BiLSTM3-256–16f). 10ms-long ($w = 10$) rectangular window, 900 time steps ($ts = 900$), learning rate $lr = 0.0005$, and mini-batch size $bs = 16$ were the best options. Finally, this model (hereinafter referred to simply as BiLSTM) was finalized, i.e., trained both on train and validation datasets, and evaluated on the evaluation datasets in Sect. 4.2. For the experiments with a joint CNN-RNN architecture in Sect. 3.3, the best model on 8 kHz frame-based speech input, i.e. 3-layered BiLSTMs with 64 cells in each layer was used as well. The best setting for this model was to use 16ms-long rectangular window ($w = 16$), $ts = 700$, $lr = 0.001$, $bs = 32$.

As for the joint CNN-RNN architecture (with the best RNNs for each sampling frequency) described in Sect. 3.3, the best results were achieved for the simple architecture (containing significantly fewer parameters than other models) with 3 convolutional blocks (CNN3) and 16 kHz speech input (see CNN3-16 in Table 2). The best setting found was as follows: the number of filters in the blocks 16, 32, 64, the kernel size 7 with the stride of 1, the pooling size 3, and the padding was "same" (please see e.g. [7] for a closer explanation). Again, the resulting model (referred to as CNN-BiLSTM) was finalized and evaluated on the evaluation datasets in Sect. 4.2.

4.2   Comparison of Different GCI Detection Models

In this study, we focus on comparing different approaches to GCI detection on the *same* data. Previously, various authors tended to propose and evaluate their own methods on different datasets which makes the comparison of different approaches difficult. We compared the proposed BiLSTM and CNN-BiLSTM with the convolutional-only network InceptionV3-1D [23], with a classical ("nondeep") machine learning-based algorithm XGBoost [22] and with two traditional GCI detection methods SEDREAMS [10] and DYPSA [24]. Since SEDREAMS and DYPSA estimate GCIs also during unvoiced segments, their authors recommend filtering the detected GCIs by the output of a separate voiced/unvoiced detector. We applied an F_0 contour estimated by the REAPER algorithm [3] for this purpose. There is no need to apply such post-processing on GCIs detected by machine learning-based methods since the voiced/unvoiced pattern is used internally in these methods. To obtain consistent results, the detected GCIs were shifted towards the neighboring minimum negative sample in the speech signal.

Two voices, a US male (BDL) and a US female (SLT) from the CMU ARCTIC database [2,17], were used as test material. Each voice consists of 1132 phonetically balanced utterances of total duration $\approx$54 min per voice. Additionally, KED TIMIT database [2], comprising 453 phonetically balanced utterances ($\approx$20 min.) of a US male speaker, was also used for testing. All these datasets comprise clean speech. Ground truth GCIs were detected from contemporaneous EGG recordings in the same way as described in Sect. 2.1 (again shifted towards the neighboring minimum negative sample in the speech signal)[1]. Original speech signals were downsampled to 16 kHz and checked to have the same

---

[1] The ground truth GCIs and other data relevant to the described experiments are available online [1].

Table 3. Comparison of GCI detection of the proposed BiLSTM and CNN-BiLSTM models with other models and algorithms on the evaluation datasets.

Dataset	Method	IDR (%)	MR (%)	FAR (%)	IDA (ms)	A25 (%)	E10 (%)
BDL	CNN-BiLSTM	**95.14**	2.76	2.10	0.64	98.18	**93.44**
	BiLSTM	94.49	4.49	**1.02**	0.44	98.21	92.85
	InceptionV3-1D [23]	94.34	3.99	1.67	0.53	**98.89**	93.37
	XGBoost [22]	93.85	**2.37**	3.78	**0.41**	98.34	92.36
	SEDREAMS [10]	91.80	3.03	5.16	0.45	97.37	90.02
	DYPSA [24]	89.43	4.38	6.19	0.54	97.13	86.89
SLT	CNN-BiLSTM	**97.04**	1.76	**1.20**	**0.14**	**99.78**	**96.83**
	BiLSTM	96.82	1.96	1.22	0.15	99.73	96.57
	InceptionV3-1D [23]	96.84	1.36	1.80	0.17	99.73	96.59
	XGBoost [22]	96.05	**0.57**	3.38	0.17	99.71	95.78
	SEDREAMS [10]	94.66	1.13	4.21	0.17	99.67	94.36
	DYPSA [24]	93.25	2.91	3.84	0.32	99.39	92.75
KED	CNN-BiLSTM	**96.64**	1.67	1.69	0.26	99.63	**96.29**
	BiLSTM	96.49	2.61	**0.90**	**0.22**	**99.69**	96.21
	InceptionV3-1D [23]	96.22	2.71	1.08	0.24	99.60	95.87
	XGBoost [22]	95.70	**1.29**	3.02	0.25	99.64	95.37
	SEDREAMS [10]	92.30	6.03	1.66	0.29	99.12	91.76
	DYPSA [24]	90.27	7.07	2.65	0.30	99.25	89.72
TOTAL	CNN-BiLSTM	**96.31**	2.12	1.58	0.41	99.18	**95.54**
	BiLSTM	95.93	2.95	**1.11**	**0.29**	99.18	95.18
	InceptionV3-1D [23]	95.87	2.46	1.68	0.35	**99.41**	95.35
	XGBoost [22]	95.22	**1.30**	3.48	**0.29**	99.21	94.49
	SEDREAMS [10]	93.37	2.34	4.29	0.31	98.79	92.51
	DYPSA [24]	90.27	7.07	2.65	0.30	99.25	89.72

polarity as described in Sect. 2.1. It is important to mention that none of the voices from these datasets was part of the training dataset used to train the machine-learning models.

The results in Table 3 confirm that machine learning-based algorithms clearly outperform the traditional ones for all testing datasets[2]. Deep learning approaches (CNN-BiLSTM, BiLSTM, and InceptionV3-1D) tend to perform better than non-deep XGBoost.

As for the comparison of RNN (BiLSTM) and CNN (InceptionV3-1D) GCI detection, the RNN model tends to be better in *reliability*, especially with respect to the identification (IDR) and false alarm (FAR) rates, suggesting that captur-

---

[2] A possible explanation of lower performance metrics (cf. e.g. the classic studies [11, 24]) is the use of different ground truth GCIs, a different strategy of GCI filtering in unvoiced segments, and perhaps also a different implementation of GCI computation evaluation (also available in [1]).

ing temporal dependencies by sequence-to-sequence modeling of input speech frames helps in better identification of GCIs.

The joint CNN-RNN (CNN-BiLSTM) architecture further enhances the GCI detection performance and excels in terms of IDR and the combined dynamic evaluation measure (E10). As for the *accuracy*, all three models performed comparably well with InceptionV3-1D being on average the best in terms of the smallest number of timing errors higher than 0.25 ms (A25) and BiLSTM being on average the best in terms of identification accuracy (IDA).

5   Conclusions

In this paper, we showed that framing GCI detection as a sequence-to-sequence prediction problem in which temporal dependencies could be interpreted across a sequence of speech frames leads to better GCI detection, especially with respect to the reliability measures (identification rate, IDR). Adding CNN layers on the front end (thus extracting relevant features from input speech) followed by recurrent layers with a dense layer on the output further improves the GCI detection performance. The proposed CNN-BiLSTM GCI detection model outperforms other machine learning-based models (either deep learning or classical non-deep learning ones) and also clearly outperforms traditional GCI detection algorithms on several public datasets.

The frame-based modeling, which respects temporal structure and dependencies present in speech, outperforms peak-based modeling. It is a good finding because no speech filtering and peak detection are required when processing input speech frame by frame.

References

1. Data used for CNN-BiLSTM glottal closure instant detection. https://github.com/ARTIC-TTS-experiments/2022-ANNPR
2. FestVox Speech Synthesis Databases. https://festvox.org/dbs/index.html
3. REAPER: Robust Epoch And Pitch EstimatoR.https://github.com/google/REAPER
4. Ardaillon, L., Roebel, A.: GCI detection from raw speech using a fully-convolutional network. In: IEEE International Conference on Acoustics Speech and Signal Processing, pp. 6739–6743. Barcelona, Spain (2020). https://doi.org/10.1109/ICASSP40776.2020.9053089
5. Barnard, E., Cole, R.A., Vea, M.P., Alleva, F.A.: Pitch detection with a neural-net classifier. IEEE Trans. Signal Process. **39**(2), 298–307 (1991). https://doi.org/10.1109/78.80812
6. Cho, K., et al.: Learning phrase representations using RNN encoder-decoder for statistical machine translation. In: Conference on Empirical Methods in Natural Language Processing, pp. 1724–1734. Doha, Qatar (2014)
7. Dhillon, A., Verma, G.K.: Convolutional neural network: a review of models, methodologies and applications to object detection. Prog. Artif. Intell. **9**(2), 85–112 (2019). https://doi.org/10.1007/s13748-019-00203-0

8. Donahue, J., et al.: Long-term recurrent convolutional networks for visual recognition and description. IEEE Trans. Pattern Anal. Mach. Intell. **39**(4), 677–691 (2015). https://doi.org/10.1109/TPAMI.2016.2599174

9. Drugman, T., Alku, P., Alwan, A., Yegnanarayana, B.: Glottal source processing: from analysis to applications. Comput. Speech Lang. **28**(5), 1117–1138 (2014). https://doi.org/10.1016/j.csl.2014.03.003

10. Drugman, T., Dutoit, T.: Glottal closure and opening instant detection from speech signals. In: INTERSPEECH, pp. 2891–2894. Brighton, Great Britain (2009)

11. Drugman, T., Thomas, M., Gudnason, J., Naylor, P., Dutoit, T.: Detection of glottal closure instants from speech signals: a quantitative review. IEEE Trans. Audio Speech Lang. Process. **20**(3), 994–1006 (2012). https://doi.org/10.1109/TASL.2011.2170835

12. Goyal, M., Srivastava, V., Prathosh, A.P.: Detection of glottal closure instants from raw speech using convolutional neural networks. In: INTERSPEECH, pp. 1591–1595. Graz, Austria (2019). https://doi.org/10.21437/Interspeech.2019--2587

13. Hochreiter, S., Schmidhuber, J.: Long Short-Term Memory. Neural Comput. **9**(8), 1735–1780 (1997). https://doi.org/10.1162/neco.1997.9.8.1735

14. Hussain, M.S., Haque, M.A.: SwishNet: a fast convolutional neural network for speech, music and noise classification and segmentation (2018). https://arxiv.org/abs/1812.00149

15. Khanagha, V., Daoudi, K., Yahia, H.M.: Detection of glottal closure instants based on the microcanonical multiscale formalism. IEEE/ACM Trans. Audio Speech Lang. Process. **22**(12), 1941–1950 (2014). https://doi.org/10.1109/TASLP.2014.2352451

16. Kiranyaz, S., Ince, T., Abdeljaber, O., Avci, O., Gabbouj, M.: 1-D convolutional neural networks for signal processing applications. In: IEEE International Conference on Acoustics Speech and Signal Processing, pp. 8360–8363. Brighton, United Kingdom (2019). https://doi.org/10.1109/ICASSP.2019.8682194

17. Kominek, J., Black, A.W.: The CMU ARCTIC speech databases. In: Speech Synthesis Workshop, pp. 223–224. Pittsburgh, USA (2004)

18. Legát, M., Matoušek, J., Tihelka, D.: A robust multi-phase pitch-mark detection algorithm. In: INTERSPEECH. vol. 1, pp. 1641–1644. Antwerp, Belgium (2007)

19. Legát, M., Matoušek, J., Tihelka, D.: On the detection of pitch marks using a robust multi-phase algorithm. Speech Commun. **53**(4), 552–566 (2011). https://doi.org/10.1016/j.specom.2011.01.008

20. Legát, M., Tihelka, D., Matoušek, J.: Pitch marks at peaks or valleys? In: Matoušek, V., Mautner, P. (eds.) TSD 2007. LNCS (LNAI), vol. 4629, pp. 502–507. Springer, Heidelberg (2007). https://doi.org/10.1007/978-3-540-74628-7_65

21. Matoušek, J., Tihelka, D.: Classification-based detection of glottal closure instants from speech signals. In: INTERSPEECH, pp. 3053–3057. Stockholm, Sweden (2017). https://doi.org/10.21437/Interspeech.2017--213

22. Matoušek, J., Tihelka, D.: Using extreme gradient boosting to detect glottal closure instants in speech signal. In: IEEE International Conference on Acoustics Speech and Signal Processing, pp. 6515–6519. Brighton, United Kingdom (2019). https://doi.org/10.1109/ICASSP.2019.8683889

23. Matoušek, J., Tihelka, D.: A comparison of convolutional neural networks for glottal closure instant detection from raw speech. In: IEEE International Conference on Acoustics Speech and Signal Processing. pp. 6938–6942. Toronto, Canada (2021). https://doi.org/10.1109/ICASSP39728.2021.9413675

24. Naylor, P.A., Kounoudes, A., Gudnason, J., Brookes, M.: Estimation of glottal closure instants in voiced speech using the DYPSA algorithm. IEEE Trans. Audio Speech Lang. Process. **15**(1), 34–43 (2007). https://doi.org/10.1109/TASL.2006. 876878

25. Prathosh, A.P., Ananthapadmanabha, T.V., Ramakrishnan, A.G.: Epoch extraction based on integrated linear prediction residual using plosion index. IEEE Trans. Audio Speech Lang. Process. **21**(12), 2471–2480 (2013). https://doi.org/10.1109/ TASL.2013.2273717

26. Purwins, H., Li, B., Virtanen, T., Schl, J., Chang, S.-Y., Sainath, T.: Deep learning for audio signal processing. IEEE J. Selected Top. Sig. Process. **13**(2), 206–219 (2019). https://doi.org/10.1109/JSTSP.2019.2908700

27. Reddy, G.M., Rao, K.S., Das, P.P.: Glottal closure instants detection from speech signal by deep features extracted from raw speech and linear prediction residual. In: INTERSPEECH, pp. 156–160. Graz, Austria (2019)

28. Schmidhuber, J.: Deep Learning in neural networks: an overview. Neural Netw. **61**, 85–117 (2015). https://doi.org/10.1016/j.neunet.2014.09.003

29. Simonyan, K., Zisserman, A.: Very deep convolutional networks for large-scale image recognition. In: International Conference on Learning Representations. San Diego, USA (2015)

30. Socher, R., Lin, C.C.-Y., Ng, A.Y., Manning, C.D.: Parsing natural scenes and natural language with recursive neural networks. In: International Conference on Machine Learning, pp. 129–136. Bellevue, Washington, USA (2011)

31. Steiner, P., Howard, I.S., Birkholz, P.: Glottal closure instance detection using Echo State Networks. In: Studientexte zur Sprachkommunikation: Elektronische Sprachsignalverarbeitung, pp. 161–168. Berlin, Germany (2021)

32. Sujith, P., Prathosh, A.P., Ramakrishnan, A.G., Ghosh, P.K.: An error correction scheme for GCI detection algorithms using pitch smoothness criterion. In: INTER-SPEECH. pp. 3284–3288. Dresden, Germany (2015)

33. Thomas, M.R.P., Gudnason, J., Naylor, P.A.: Data-driven voice source waveform modelling. In: IEEE International Conference on Acoustics Speech and Signal Processing, pp. 3965–3968. Taipei, Taiwan (2009). https://doi.org/10.1109/ICASSP. 2009.4960496

34. Thomas, M.R.P., Gudnason, J., Naylor, P.A.: Estimation of glottal closing and opening instants in voiced speech using the YAGA algorithm. IEEE Trans. Audio Speech Lang. Process. **20**(1), 82–91 (2012). https://doi.org/10.1109/TASL.2011. 2157684

35. Tuan, V.N., D'Alessandro, C.: Robust glottal closure detection using the wavelet transform. In: EUROSPEECH, pp. 2805–2808. Budapest, Hungary (1999)

Mono vs Multilingual BERT for Hate Speech Detection and Text Classification: A Case Study in Marathi

Abhishek Velankar[1,3], Hrushikesh Patil[1,3(✉)], and Raviraj Joshi[2,3]

[1] Pune Institute of Computer Technology, Pune, Maharashtra, India
[2] Indian Institute of Technology Madras, Chennai, Tamilnadu, India
[3] L3Cube, Pune, India
hrushi2900@gmail.com

Abstract. Transformers are the most eminent architectures used for a vast range of Natural Language Processing tasks. These models are pre-trained over a large text corpus and are meant to serve state-of-the-art results over tasks like text classification. In this work, we conduct a comparative study between monolingual and multilingual BERT models. We focus on the Marathi language and evaluate the models on the datasets for hate speech detection, sentiment analysis, and simple text classification in Marathi. We use standard multilingual models such as mBERT, indicBERT, and xlm-RoBERTa and compare them with MahaBERT, MahaALBERT, and MahaRoBERTa, the monolingual models for Marathi. We further show that Marathi monolingual models outperform the multilingual BERT variants in five different downstream fine-tuning experiments. We also evaluate sentence embeddings from these models by freezing the BERT encoder layers. We show that monolingual MahaBERT-based models provide rich representations as compared to sentence embeddings from multi-lingual counterparts. However, we observe that these embeddings are not generic enough and do not work well on out-of-domain social media datasets. We consider two Marathi hate speech datasets L3Cube-MahaHate, HASOC-2021, a Marathi sentiment classification dataset L3Cube-MahaSent, and Marathi Headline, Articles classification datasets.

Keywords: Natural language processing · Text classification · Hate speech detection · Sentiment analysis · BERT · Marathi BERT

1 Introduction

The language models like BERT, built over the transformer architecture, have gained a lot of popularity due to the promising results on an extensive range of natural language processing tasks. These large models make use of attention mechanism from transformers and understand the language deeper in terms of context. These models can be fine-tuned on domain-specific data to obtain state-of-the-art solutions.

© The Author(s), under exclusive license to Springer Nature Switzerland AG 2023
N. El Gayar et al. (Eds.): ANNPR 2022, LNAI 13739, pp. 121–128, 2023.
https://doi.org/10.1007/978-3-031-20650-4_10

More recently, there has been a significant amount of research on monolingual and multilingual language models, specifically the BERT variants. Due to the variety of text corpus in terms of languages used for training, multilingual models find notable benefits over multiple applications, specifically for languages that are low in resources [8, 19, 20]. However, the monolingual models, when used on the corresponding language, outperform the multilingual versions in tasks like text classification. Both the categories of models find their use in several problems like next sentence prediction, named entity recognition, sentiment analysis, etc. Recently, a substantial amount of work can be seen with the use of these models on native languages. [3] propose monolingual BERT models for the Arabic language and show that these models achieve state-of-the-art performances. Additionally, [14, 25, 26] show that the single language models, when used for the corresponding language tasks, perform more efficiently than the multilingual variants. [21] analyze the effectiveness of multilingual models over their monolingual counterparts for 6 different languages including English and German. Our work focuses on hate speech detection, sentiment analysis, and simple text classification in Marathi [12, 13, 27]. We evaluate monolingual and multilingual BERT models on the Marathi corpus to compare their performance. A similar analysis for Hindi and Marathi named entity recognition has been performed in [15].

Marathi is a regional language in India. It is majorly spoken by the people in Maharashtra [9]. Additionally, after Hindi and Bengali, it is considered as the third most popular language in India [4, 7]. However, the Marathi language is greatly overlooked in terms of language resources which suggests the need of widening the research in this area [10].

In this work, we perform a comparative analysis of monolingual and multilingual BERT models for Marathi. We fine-tune these models over the Marathi corpus, which contains hate speech detection and simple text classification datasets. We consider standard multilingual models i.e. mBERT, indicBERT and xlm-RoBERTa and compare them with Marathi monolingual counterparts i.e. MahaBERT, MahaALBERT and MahaRoBERTa. We further show that the monolingual models when used on Marathi, outperform the multilingual equivalents. Moreover, we evaluate sentence representations from these models and show that the monolingual models provide superior sentence representations. The advantage of using monolingual models is more visible when extracted sentence embeddings are used for classification. This research is aimed to help the community by giving an insight into the appropriate use of these single and multilingual models when applied to single language tasks.

2   Related Work

The BERT is currently one of the most effective language models in terms of performance when different NLP tasks like text classification are concerned. Previous research has shown how BERT captures the language context efficiently [6, 24, 29].

Recently, a lot of work can be seen in single and multi-language NLP applications. Several efforts have been made to build monolingual variants of BERT and shown to be effective over a quantity of single language downstream tasks. In [22] authors publish a German monolingual BERT model based on RoBERTa. The experiments have been performed on the tasks like named entity recognition (NER) and text classification to evaluate the model performance. They further propose that, with only little tuning of hyperparameters, the model outperformed all other tested German and multilingual BERT models. A monolingual RoBERTa language model trained on Czech data has been presented in [23]. Authors show that the model significantly outperforms equally-sized multilingual and Czech language-oriented model variants. Other works for single language-specific BERT models include models built in Vietnamese, Hindi, Bengali, etc. [5,17]. In [18] authors propose model evaluations on toxicity detection in Spanish comments. They show that transformers obtain better results than statistical models. Furthermore, they conclude monolingual BERT models provide better results in their pre-trained language as compared to multilingual models.

3   Datasets

- **HASOC'21 Marathi dataset** [16]:
  Marathi binary dataset provided in HASOC'21 shared task divided into hateful and non-hateful categories. It consists of a total of 1874 training and 625 testing samples.
- **L3Cube-MahaHate** [28]:
  hate speech detection dataset in Marathi consisting of 25000 tweet samples divided into 4 major classes namely hate, offensive, profane and not. The dataset consists of 21500 train, 2000 test and 1500 validation examples.
- **Articles:**
  text classification dataset containing Marathi news articles classified into sports, entertainment, and lifestyle with 3823 train, 479 test, and 477 validation samples.
- **Headlines:**
  Marathi news headlines dataset containing the headlines containing three classes viz. entertainment, sports, and state. It consists of 9672 train, 1210 test and 1210 validation samples.
- **L3Cube-MahaSent** [13]:
  Sentiment Analysis dataset in Marathi includes tweets classified as positive, negative, and neutral. It has 12114 train, 2250 test, and 1500 validation examples.

4   Experiments

4.1   Transformer Models

BERT is a deep transformer model, pre-trained over a large text corpus in a self-supervised manner and provides a great ability to promptly adapt to a broad range of downstream tasks. There are a lot of different flavors of BERT available openly, some popular variants are ALBERT and RoBERTa. In this work, we are focusing on both multilingual and monolingual models for text classification and hate speech detection tasks. Following standard multilingual BERT models which use Marathi as one of the training languages are used:

- **Multilingual-BERT (mBERT)**[1]: It is a BERT-base model [2] trained on and usable with 104 languages with Wikipedia using a masked language modeling (MLM) and next sentence prediction (NSP) objective.
- **IndicBERT**[2]: a multilingual ALBERT model released by Ai4Bharat [11], trained on large-scale corpora. The training languages include 12 major Indian languages. The model has been proven to be working better for tasks in Indic languages.
- **XLM-RoBERTa**[3]: a multilingual version of the RoBERTa model [1]. It is pre-trained on 2.5TB of filtered CommonCrawl data containing 100 languages with the Masked language modeling (MLM) objective and can be used for downstream tasks.

To compare with the above models, the following Marathi monolingual models are used [9]:

- **MahaBERT**[4]: a multilingual BERT model fine-tuned on L3Cube-MahaCorpus and other publicly available Marathi monolingual datasets containing a total of 752M tokens.
- **MahaAlBERT**[5]: It is a Marathi monolingual model extended from AlBERT, trained on L3Cube-MahaCorpus and other publicly available Marathi monolingual datasets.
- **MahaRoBERTa**[6]: It is a Marathi RoBERTa model built upon a multilingual RoBERTa (xlm-roberta-base) model fine-tuned on L3Cube-MahaCorpus and other publicly available Marathi monolingual datasets.

---

[1] https://huggingface.co/bert-base-multilingual-cased.
[2] https://huggingface.co/ai4bharat/indic-bert.
[3] https://huggingface.co/xlm-roberta-base.
[4] https://huggingface.co/l3cube-pune/marathi-bert.
[5] https://huggingface.co/l3cube-pune/marathi-albert.
[6] https://huggingface.co/l3cube-pune/marathi-roberta.

Table 1. Classification accuracies for monolingual and multilingual models.

Model	Training mode	HASOC	L3Cube-MahaHate	L3Cube-MahaSent	Articles	Headlines
Multilingual BERT Variants						
mBERT	Freeze	0.770	0.516	0.653	0.901	0.907
	Non-Freeze	**0.875**	0.783	0.786	0.976	**0.947**
IndicBERT	Freeze	0.710	0.436	0.656	0.828	0.877
	Non-Freeze	0.870	0.711.	**0.833**	**0.987**	0.937
xlm-RoBERTa	Freeze	0.755	0.487	0.666	0.91	0.79
	Non-Freeze	0.862	**0.787**	0.820	0.985	0.925
Monolingual BERT Variants						
MahaBERT	Freeze	0.824	0.580	0.666	0.939	0.907
	Non-Freeze	0.883	0.802	0.828	0.987	0.944
MahaAlBERT	Freeze	0.800	0.587	0.717	0.991	0.927
	Non-Freeze	0.866	0.764	**0.841**	**0.991**	**0.945**
MahaRoBERTa	Freeze	0.782	0.531	0.698	0.904	0.864
	Non-Freeze	**0.890**	**0.803**	0.834	0.985	0.942

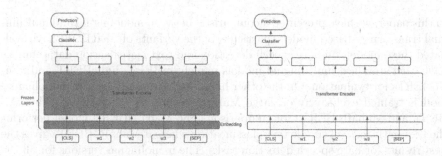

Fig. 1. BERT architectures with freeze and non-freeze training mode

4.2 Evaluation Results

The BERT transformer models have been evaluated on hate speech detection and text classification datasets. We used standard multilingual BERT variants namely indicBERT, mBERT and xlm-RoBERTa to obtain baseline classification results. Additionally, monolingual Marathi models have been used for comparison. These single language models include MahaBERT, MahaAlBERT and MahaRoBERTa are based on the BERT-base, AlBERT and RoBERTa-base models respectively.

The experiments have been performed in two schemes. Firstly, we obtained the results by fine-tuning all the BERT layers i.e. pre-trained layers as well as classification layers. Furthermore, we froze the pre-trained embedding and encoder layers and trained only the classifier to obtain the results. Using this setup we aim to evaluate the sentence embeddings generated by these monolingual and multilingual models. All the classification results are displayed in Table 1.

For all the monolingual and multilingual models, the frozen settings i.e. freezing BERT embedding and encoder layers are underperforming as compared to their non-freeze counterparts. The difference in accuracy is too high for L3Cube-MahaSent and L3Cube-MahaHate. This indicates that the pre-trained models do not provide generic discriminative sentence embeddings for the classification task. However, the mono-lingual model does provide better sentence embeddings as compared to the multi-lingual counterpart. This shows the importance of monolingual pretraining for obtaining rich sentence embeddings. Since the pre-training data mostly comprised of Marathi news articles the frozen setting works comparatively well on the Articles and Headlines dataset. In general, the monolingual models have outperformed the multilingual models on all the datasets. For hate speech detection datasets, particularly the MahaRoBERTa model is working the best. In the case of other text classification datasets, the MahaAlBERT model is giving the best accuracy.

5   Conclusion

In this paper, we have presented a comparison between monolingual and multilingual transformer-based models, particularly the variants of BERT. We have evaluated these models on hate speech detection and text classification datasets. We have used standard multilingual models namely mBERT, indicBERT and xlm-RoBERTa for evaluation. On the other hand, we have used Marathi monolingual models trained exclusively on large Marathi corpus i.e. MahaBERT, MahaAlBERT and MahaRoBERTa for comparison. The MahaAlBERT model performs the best in the case of simple text classification whereas MahaRoBERTa gives the best results for hate speech detection tasks. The monolingual versions for all the datasets have outperformed the standard multilingual models when focused on single language tasks. The monolingual models also provide better sentence representations. However, these sentence representations do not generalize well across the tasks, thus highlighting the need for better sentence embedding models.

Acknowledgements. This work was done under the L3Cube Pune mentorship program. We would like to express our gratitude towards our mentors at L3Cube for their continuous support and encouragement.

References

1. Conneau, A., et al.: Unsupervised cross-lingual representation learning at scale. CoRR abs/1911.02116 (2019). https://arxiv.org/abs/1911.02116
2. Devlin, J., Chang, M., Lee, K., Toutanova, K.: BERT: pre-training of deep bidirectional transformers for language understanding. CoRR abs/1810.04805 (2018). https://arxiv.org/abs/1810.04805
3. Ghaddar, A., et al.: JABER: junior arabic bert. CoRR abs/2112.04329 (2021). https://arxiv.org/abs/2112.04329

4. Islam, M.S., Jubayer, F.E.M., Ahmed, S.I.: A comparative study on different types of approaches to Bengali document categorization. CoRR abs/1701.08694 (2017). https://arxiv.org/abs/1701.08694
5. Jain, K., Deshpande, A., Shridhar, K., Laumann, F., Dash, A.: Indic-transformers: an analysis of transformer language models for Indian languages. CoRR abs/2011.02323 (2020). https://arxiv.org/abs/2011.02323
6. Jawahar, G., Sagot, B., Seddah, D.: What does BERT learn about the structure of language? In: Proceedings of the 57th Annual Meeting of the Association for Computational Linguistics, pp. 3651–3657. Association for Computational Linguistics, Florence, Italy (2019). https://doi.org/10.18653/v1/P19-1356. https://aclanthology.org/P19-1356
7. Joshi, R., Goel, P., Joshi, R.: Deep learning for Hindi text classification: a comparison. In: Tiwary, U.S., Chaudhury, S. (eds.) IHCI 2019. LNCS, vol. 11886, pp. 94–101. Springer, Cham (2020). https://doi.org/10.1007/978-3-030-44689-5_9
8. Joshi, R., Karnavat, R., Jirapure, K., Joshi, R.: Evaluation of deep learning models for hostility detection in Hindi text. In: 2021 6th International Conference for Convergence in Technology (I2CT), pp. 1–5. IEEE (2021)
9. Joshi, R.: L3cube-mahacorpus and mahabert: Marathi monolingual corpus, marathi BERT language models, and resources. CoRR abs/2202.01159 (2022). https://arxiv.org/abs/2202.01159
10. Joshi, R.: L3cube-mahanlp: Marathi natural language processing datasets, models, and library. arXiv preprint arXiv:2205.14728 (2022)
11. Kakwani, D., et al.: IndicNLPSuite: monolingual corpora, evaluation benchmarks and pre-trained multilingual language models for indian languages. In: Findings of EMNLP (2020)
12. Kulkarni, A., Mandhane, M., Likhitkar, M., Kshirsagar, G., Jagdale, J., Joshi, R.: Experimental evaluation of deep learning models for marathi text classification. In: Gunjan, V.K., Zurada, J.M. (eds.) Proceedings of the 2nd International Conference on Recent Trends in Machine Learning, IoT, Smart Cities and Applications. LNNS, vol. 237, pp. 605–613. Springer, Singapore (2022). https://doi.org/10.1007/978-981-16-6407-6_53
13. Kulkarni, A., Mandhane, M., Likhitkar, M., Kshirsagar, G., Joshi, R.: L3cubemahasent: a Marathi tweet-based sentiment analysis dataset. In: Clercq, O.D., et al. (eds.) Proceedings of the Eleventh Workshop on Computational Approaches to Subjectivity, Sentiment and Social Media Analysis, WASSA@EACL 2021, Online, 19 April 2021, pp. 213–220. Association for Computational Linguistics (2021). https://aclanthology.org/2021.wassa-1.23/
14. Le, H., et al.: FlauBERT: unsupervised language model pre-training for French. In: Proceedings of the 12th Language Resources and Evaluation Conference, pp. 2479–2490. European Language Resources Association, Marseille, France (2020). https://aclanthology.org/2020.lrec-1.302
15. Litake, O., Sabane, M., Patil, P., Ranade, A., Joshi, R.: Mono vs multilingual BERT: a case study in Hindi and Marathi named entity recognition. CoRR abs/2203.12907 (2022). https://doi.org/10.48550/arXiv.2203.12907
16. Modha, S., et al.: Overview of the hasoc subtrack at fire 2021: hate speech and offensive content identification in English and Indo-Aryan languages and conversational hate speech. In: Forum for Information Retrieval Evaluation, pp. 1–3 (2021)
17. Nguyen, D.Q., Nguyen, A.T.: Phobert: pre-trained language models for vietnamese. CoRR abs/2003.00744 (2020). https://arxiv.org/abs/2003.00744

18. de Paula, A.F.M., Schlicht, I.B.: AI-UPV at iberlef-2021 DETOXIS task: toxicity detection in immigration-related web news comments using transformers and statistical models. CoRR abs/2111.04530 (2021). https://arxiv.org/abs/2111.04530

19. Pires, T., Schlinger, E., Garrette, D.: How multilingual is multilingual bert? In: Korhonen, A., Traum, D.R., Màrquez, L. (eds.) Proceedings of the 57th Conference of the Association for Computational Linguistics, ACL 2019, Florence, Italy, 28 July – 2 August 2019, vol. 1: Long Papers, pp. 4996–5001. Association for Computational Linguistics (2019). https://doi.org/10.18653/v1/p19-1493

20. Pires, T., Schlinger, E., Garrette, D.: How multilingual is multilingual BERT? In: Proceedings of the 57th Annual Meeting of the Association for Computational Linguistics, pp. 4996–5001. Association for Computational Linguistics, Florence, Italy (2019). https://doi.org/10.18653/v1/P19-1493. https://aclanthology.org/P19-1493

21. Rönnqvist, S., Kanerva, J., Salakoski, T., Ginter, F.: Is multilingual BERT fluent in language generation? CoRR abs/1910.03806 (2019). https://arxiv.org/abs/1910.03806

22. Scheible, R., Thomczyk, F., Tippmann, P., Jaravine, V., Boeker, M.: Gottbert: a pure German language model. CoRR abs/2012.02110 (2020). https://arxiv.org/abs/2012.02110

23. Straka, M., Náplava, J., Straková, J., Samuel, D.: Robeczech: Czech roberta, a monolingual contextualized language representation model. CoRR abs/2105.11314 (2021). https://arxiv.org/abs/2105.11314

24. Tenney, I., Das, D., Pavlick, E.: BERT rediscovers the classical NLP pipeline. In: Proceedings of the 57th Annual Meeting of the Association for Computational Linguistics, pp. 4593–4601. Association for Computational Linguistics, Florence, Italy (2019). https://doi.org/10.18653/v1/P19-1452. https://aclanthology.org/P19-1452

25. To, H.Q., Nguyen, K.V., Nguyen, N.L., Nguyen, A.G.: Monolingual versus multilingual bertology for vietnamese extractive multi-document summarization. CoRR abs/2108.13741 (2021). https://arxiv.org/abs/2108.13741

26. Ulcar, M., Robnik-Sikonja, M.: Training dataset and dictionary sizes matter in BERT models: the case of baltic languages. CoRR abs/2112.10553 (2021). https://arxiv.org/abs/2112.10553

27. Velankar, A., Patil, H., Gore, A., Salunke, S., Joshi, R.: Hate and offensive speech detection in hindi and marathi. CoRR abs/2110.12200 (2021). https://arxiv.org/abs/2110.12200

28. Velankar, A., Patil, H., Gore, A., Salunke, S., Joshi, R.: L3cube-mahahate: a tweet-based marathi hate speech detection dataset and BERT models. CoRR abs/2203.13778 (2022). https://doi.org/10.48550/arXiv.2203.13778

29. de Vries, W., van Cranenburgh, A., Nissim, M.: What's so special about BERT's layers? a closer look at the NLP pipeline in monolingual and multilingual models. In: Findings of the Association for Computational Linguistics: EMNLP 2020, pp. 4339–4350. Association for Computational Linguistics, Online (Nov 2020). https://doi.org/10.18653/v1/2020.findings-emnlp.389. https://aclanthology.org/2020.findings-emnlp.389

Transformer-Encoder Generated Context-Aware Embeddings for Spell Correction

Noufal Samsudin[1]([✉]) [iD] and Hani Ragab Hassen[2] [iD]

[1] Dubai Holding, Dubai, United Arab Emirates
noufal.samsudin@dubaiholding.com
[2] Heriot-Watt University, Dubai, United Arab Emirates
h.ragabhassen@hw.ac.uk

Abstract. In this paper, we propose a novel approach for context-aware spell correction in text documents. We present a deep learning model that learns a context-aware character-level mapping of words to a compact embedding space. In the embedding space, a word and its spelling variations are mapped close to each other in a Euclidean space. After we develop this mapping for all words in the dataset's vocabulary, it is possible to identify and correct wrongly spelt words by comparing the distances of their mappings with those of the correctly spelt words. The word embeddings are built in a way that captures the context of each word. This makes it easier for our system to identify correctly spelt words that are used out of their contexts (e.g., their/there, your/you're). The Euclidean distance, between our word embeddings, can thus be deemed as a context-aware string similarity metric.

We employ a transformer-encoder model that takes character-level input of words and their context to achieve this. The embeddings are generated as the outputs of the model. The model is then trained to minimize triplet loss, which ensures that spell variants of a word are embedded close to the word, and that unrelated words are embedded farther away. We further improve the efficiency of the training by using a hard triplet mining approach. Our approach was inspired by FaceNet [18], where the authors developed a similar approach for face recognition and clustering using embeddings generated from Convolutional Neural Networks. The results of our experiments show that our approach is effective in spell check applications.

Keywords: Spell correction · Multi-headed attention · Transformer · Triplet loss

1 Introduction

Spelling mistakes affect computers far more than humans. Humans can easily comprehend text with spelling mistakes because of our remarkable ability to understand their context. Often, we do not even notice the spelling mistakes.

Computers on the other hand are not as efficient in text comprehension. Spelling errors can significantly compromise the effectiveness of document search and retrieval algorithms.

Spelling errors can be of 2 types: non-word error and real-word error. Non-word errors are errors where the misspelled word does not belong to the language's word corpus. For example, "cement" misspelled as "sement", "chicken"-"chiken". Such errors are relatively easier to correct algorithmically by relying on string distance metrics to compare these words to the word corpus. The distance metrics are usually Levenshtein Distance [10] or n-gram based methods [3].

Real-word errors on the other hand, are more difficult to detect. These are words that appear in the dictionary but are erroneous within the context. E.g. in the text "The men power required for this job...", "man" in "man power" has been misspelt as "men". Since "men" also appears in the dictionary, information of the context is critical in detecting and rectifying this error. Traditionally, context-aware spell checkers rely on building n-gram statistics for words in a corpus. This is then be used to calculate the likelihood that a word "belongs" in a particular context.

In this paper we propose a deep learning based character-level word embedding approach, which also captures the context of a word. We generate embeddings for all words in the vocabulary and index them. Our spell check algorithm generates the embeddings of the query word-context pair and calculates its Euclidean distance to the indexed embeddings to find the closest match. If the word is misspelt in that context, then the correct spelling of the word would be retrieved. We would use a Transformer-encoder model trained on triplet loss function to generate the embeddings. This approach was inspired by FaceNet [18], where the authors embed images of faces in an embedding space and rely on a similar technique for face recognition. We evaluate the performance of our approach on 2 datasets and benchmark it with other spell-correction techniques.

2   Related Work

Most traditional spell checkers employ some variant of the below approach [3] [15]: (1) Generate candidates for corrections using some noisy channel model-based on some heuristics like spell variant rules, edits, scrambling etc. (2) Score them based on probability scores. The probability scores are estimated from pre-calculated n-gram statistics. The n-gram statistics are calculated on how frequently a sequence appears in the corpus.

Carlson and Fette [1] further extended this approach and tested a variation of this approach on Google n-gram corpus of 10^9 words and reported high accuracy on the same.

2.1   Deep Learning Based Approaches to Spell Correction

The most common approach to spell correction using neural network is using a sequence to sequence model, that takes the character sequence of the word/sentence as input and generate the correct sequence of words in the corpus as output. The context can also be provided as input to the neural network by providing the entire sentence as the input in one setting.

This was achieved using models like LSTMs and GRUs which can take sequences as input and generating sequences as output. Semi-character recurrent neural network (scRNN) [17] is an example of such a model using LSTMs, which takes a noisy character level sequence as input and predicts the correct word as output. LSTM-Char-CNN [8] uses a similar approach where the character level input is processed by a CNN and the output of the CNN is fed into a LSTM, which in turn predicts the correct word as the output. Similarly, nested RNNs [12] , GRU (CharRNN) [5] units to process character level data and the sequence is in turn fed into another set of GRUs (wordRNN).

The transformer architecture is increasingly being used for several natural language processing tasks, including classification, text generation, text summarization, and translation [9,11,20,21] . It is being deemed as a scalable alternative to recurrence for sequence modelling [20]. The transformer architecture [19] has 2 main components - an encoder (that processes sequential input information) and a decoder that can generate sequential output. Transformers rely on multi-head self-attention mechanism as a substitute for recurrence in RNN based models. Hu et al. [6] used a BERT model to generate candidates for a masked input and determines ranks them by edit distance from the masked word. "Context-aware Stand-alone Neural Spelling Correction" [13] explored training a word+char transformer-encoder on noisy character level inputs to predict the correct word as output.

Triplet loss, the loss function used in this paper is primarily used for image retrieval and face recognition applications [18]. We are exploring using this function for spell check application.

3   Proposed Method

We propose to use a transformer-encoder model that takes character-level input. The model input is a combination of the word and its context as tokenized character sequence. The model outputs a fixed dimension embedding. The model is trained to minimize triplet loss - this means that the embeddings generated from the word-context pair input, would be such that the distances between a word and its corresponding spelling variations would be smaller than distances between different words. Once the model has been trained to generate these embeddings, we would then generate embeddings for all the correct words-context pairs in the dictionary and index these embeddings. To perform spell check, we generate embeddings for the query word along with its context and look up the word in the indexed embeddings to find the corrected word. Our Github repository contains the source code for our method[1].

---

[1] https://github.com/kvsnoufal/wordFaceNet.

3.1   Model Architecture

We developed a Transformer-Encoder model to generate the embeddings. The encoder has four encoder layers stacked, each with four self-attention heads, with input embedding dimension of 200 and 256 neurons in feed-forward layer. The output of the encoder has a dimension (d) of 64. The optimal architecture was reached using Bayesian hyperparameter search.

The encoder takes three inputs (see Fig. 1):

1. Tokenized input: The word and context are padded to fixed character lengths each. They are then tokenized at character level.
2. Token Type: The token type is an indicator to help the model differentiate between the word and the context in the input tokens.
3. Padding Mask: Padding mask indicates where the sequence has been padded. It prevents the encoder layers from applying attention on padded characters.

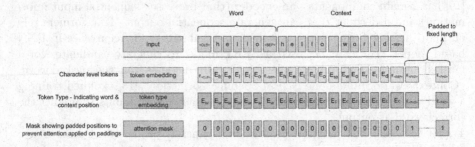

Fig. 1. Inputs to the Transformer-encoder model.

The output of the model is an embedding of dimension d followed by $L2$ normalization.

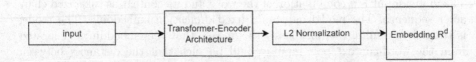

Fig. 2. The model outputs the embedding - an L2 normalized vector of dimension d

3.2   Model Training and Triplet Loss

The output of the model is a $L2$ normalized embedding of dimension d for a word-context pair input (x), represented by $f(x) \subset R^d, \|f(x)\|_2 = 1$.
Given 3 words-context inputs, namely:

1. Anchor input x_i^a with correct spelling of the word in a specific context
2. Positive input x_i^p with a misspelled variant of the anchor word and its respective context
3. Negative input x_i^n with a different word in its respective context

We need to ensure that

$$\|f(x_i^a) - f(x_i^p)\|_2^2 + \alpha < \|f(x_i^a) - f(x_i^n)\|_2^2 \tag{1}$$

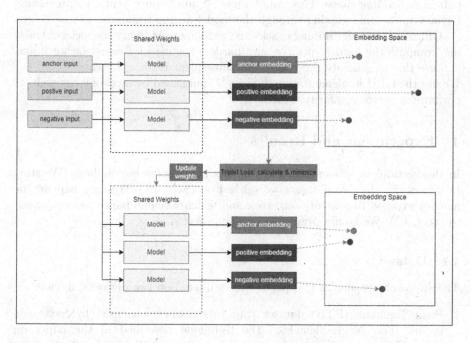

Fig. 3. Training the model would involve reducing the Triplet Loss - reducing the Euclidean distance between anchor and positive embeddings, and increasing the same between anchor and negative embeddings

This means that in the embedding space, the distance between anchor and positive inputs should be less than the distance between anchor and negative inputs by a margin alpha α (see Fig. 3). Figure 3 shows an iteration of the optimisation process where, after update the model weights, the anchor's embedding becomes closer to its variant (positive), and farther from a different word (negative). Hence, the loss function to optimize (minimize) is

$$L = \sum_i^N \left[\|f(x_i^a) - f(x_i^p)\|_2^2 - \|f(x_i^a) - f(x_i^n)\|_2^2 + \alpha \right] \tag{2}$$

An instance in the training batch for this model would comprise word-context triplets: anchor, positive and negative. The same model generates embeddings for all the triplets in a batch, and the triplet loss is calculated. It is inefficient to generate all possible triplets for training as it may increase the training time. To prevent this, we employ offline hard triplet mining described in FaceNet [18]. This means that for a given state of the model, for each anchor input, we select the negatives and positives for which the triplet loss is already high. i.e., we select hard positives; the misspellings of the anchor word but currently farther way from the anchor word-context in embedding space. We also select hard negatives; word-context which are currently closer to the anchor word-context pair in embedding space. This would allow for quick convergence, by preventing trivial triplets from passing through the model during training.

Offline hard triplet mining is achieved as follows: At Every n epochs of training, compute the hardest positive and hardest negative for every anchor input, and use this to generate the triplets for subsequent training batches.
We use the GPU implementation of Faiss [7] published by facebook research for performing vector similarity search.

4   Experiment and Results

In this section, we present our experiments and discuss our findings. We start by presenting the two datasets we worked on (Sect. 4.1). We then explain the metrics we used to evaluate our work and which baseline paper we compare it to (Sect. 4.2). We finally present our results (Sect. 4.3).

4.1   Dataset

The approach detailed in this paper was evaluated on two datasets, namely:

1. Penn Treebank (PTB) dataset [14]: this dataset was used by Sakaguchi et. al. [17]. We implemented the technique described in the paper on the dataset from the github [16] implementation. We employed the same train/validation/test split as the paper for our experiments. We used a slightly modified noise generation approach to make the noises more "realistic". We perform 4 operations to synthesis the noise in a word: (i) Swap: randomly swap adjacent characters, (ii) Drop: drop random characters, (iii) Add: add random characters, and (iv) Key: randomly swap characters that are in close proximity on a keyboard layout.
2. A corporate procurement dataset - this is a private, domain specific dataset. The raw data is a collection of typed documents. The dataset was prepared by running several string-matching algorithms and spell checking algorithms, followed by manual human validation and curating.

Table 1 gives a summary of the dataset properties.

Table 1. Dataset properties

Dataset	Number of words	Number of errors	Number of sentences/contexts		
			Training	Validation	Test
Penn Treebank [14]	9k	2k	40k	41k	2.5k
Corporate	6k	1.5k	10k	2k	1k

4.2 Training, Evaluation and Baselines

We trained the Transformer-Encoder model to generate the embeddings with Adam optimizer, with learning rate 0.0001 for 100 epochs.

We evaluate our model in comparison to scRNN [17] on the 2 datasets. The primary metric for evaluation is Accuracy. We employed a 10-fold cross validation strategy to evaluate the results. In our experiments, the models can ingest a word and producing another word as output. The output word is compared to the target. So the errors can be of two types, namely a correctly spelt word being wrongly identified as a misspelling, and a wrongly spelt word being mapped to a wrong spelling. We analyse both these errors to further understand the model behaviour. We further evaluate and compare the precision and recall as secondary metrics.

Table 2. Benchmarking performance on datasets.

Model	Penn Treebank [14]			Corporate		
	Accuracy	Precision	Recall	Accuracy	Precision	Recall
scRNN [17]	0.912	0.71	0.815	0.85	0.65	0.712
Ours[a]	**0.979**	**0.912**	**0.945**	**0.92**	**0.845**	**0.91**

[a] The metrics reported are average of 10 fold cross-validation

We used the official implementation of scRNN [16] to establish the baseline performance. The training parameters are listed in table. The implementation of scRNN on the PTB dataset employs a custom noising strategy, and so the results differ slightly from the original paper.

4.3 Results

In our experiments, our model outperformed scRNN for both the datasets (see Table 2) according to the the three metrics we are using. To better understand why it is the case, we show some examples of misspelt words and how scRNN and our approach handled them in Table 3. The embeddings generated by the model seems to be effective in spell checking. We have observed a 7.3% improvement in accuracy over scRNNs [17] in Penn Treebank dataset [14] and 8% improvement in the corporate dataset.

Table 3. Handpicked examples of model outputs.

Dataset	Type	Text
Penn Treebank	Source	buut wihle txhe new york sotck exchnge did n't fsll apart
	Target	but while the new york stock exchange did n't fall apart
	scRNN	but while the new york stocks exchange did n't fill apart
	Ours	but while the new york stock exchange did n't fall apart
Corporate	Source	provisional contact for supply of petrol & deisel fuel cards for unk security management fro the month of September & october n
	Target	provisional contract for supply of petrol & diesel fuel cards for unk security management for the months of september & october n
	scRNN	provisional contact for supply of petrol & diesel fuel cards for unk security management from the month of september & october n
	Ours	provisional contract for supply of petrol & diesel fuel cards for unk security management for the months of september & october n

We can thus infer that the embedding successfully approximates information about spelling of a word and its possible variations (This is visualized in Fig. 4- T-SNE 2d projection of the embedding space. We see that that words and their spell variants cluster close to each other, whereas different words are farther away.) We also demonstrate its effectiveness in the corporate dataset "real-world" spelling errors. This can be exploited to develop a spell checker. We believe that the improved performance over the benchmarks can be attributed to the following: (1) Multi-Headed attention mechanism in Transformer architecture is better suited at capturing sequence information than RNNs [19]). (2) scRNN's architecture and hyperparameters have been designed for synthetically generated noises and does not generalize as well in a real-world dataset.

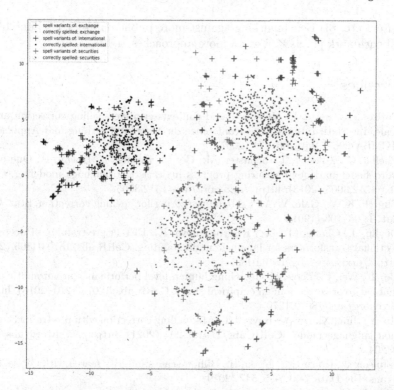

Fig. 4. T-SNE representation of embeddings generated by the model for 3 words ("international", "exchange", "securities") in different contexts and their incorrectly spelled instances in the dataset. We see that the words and their respective misspellings are embedded close to each other.

5   Conclusion and Future Work

In this work, we trained a transformer-encoder model for generating embeddings for words-context pairs, and demonstrated an approach of using these embeddings for spell checking application. We evaluated this approach on a public and a private dataset; and benchmarked its performance against the approach proposed in scRNNs paper. Our experiments showed that our approach outperforms scRNNs in both the datasets. We conclude that our approach has potential for use as spell checker and hope this stimulates further research in the field.

The scope of this paper was limited by time, cost and computational resources we had at our disposal. Future work will focus on optimizing the training process, model architecture and reducing the time for training and inferencing.

In the future we would like to investigate the following: (1) Augmenting training to generate noise at run time- similar to approach used in Sakaguchi et. al. [17] (2) Experimenting with different input formatting. E.g., providing context as word tokens instead of character tokens, masking the word position in context etc. (3) Benchmarking against more public datasets, e.g., one billion word benchmark [2], as well as on more approaches [4,12].

References

1. Carlson, A., Fette, I.: Memory-based context-sensitive spelling correction at web scale. In: Sixth International Conference on Machine Learning and Applications (ICMLA 2007), pp. 166–171 (2007)
2. Chelba, C., Mikolov, T., Schuster, M., Ge, Q., Brants, T., Koehn, P.: One billion word benchmark for measuring progress in statistical language modeling. CoRR abs/1312.3005 (2013). http://arxiv.org/abs/1312.3005
3. Church, K.W., Gale, W.A.: Probability scoring for spelling correction. Stat. Comput. **1**, 93–103 (1991)
4. Devlin, J., Chang, M., Lee, K., Toutanova, K.: BERT: pre-training of deep bidirectional transformers for language understanding. CoRR abs/1810.04805 (2018). http://arxiv.org/abs/1810.04805
5. Ge, T., Wei, F., Zhou, M.: Reaching human-level performance in automatic grammatical error correction: an empirical study. CoRR abs/1807.01270 (2018). http://arxiv.org/abs/1807.01270
6. Hu, Y., Jing, X., Ko, Y., Rayz, J.T.: Misspelling correction with pre-trained contextual language model. CoRR abs/2101.03204 (2021). https://arxiv.org/abs/2101.03204
7. Johnson, J., Douze, M., Jégou, H.: Billion-scale similarity search with GPUs. IEEE Trans. Big Data **7**(3), 535–547 (2019)
8. Kim, Y., Jernite, Y., Sontag, D., Rush, A.M.: Character-aware neural language models. In: Proceedings of the Thirtieth AAAI Conference on Artificial Intelligence, pp. 2741–2749. AAAI2016, AAAI Press (2016)
9. Lample, G., Conneau, A.: Cross-lingual language model pretraining. CoRR abs/1901.07291 (2019). http://arxiv.org/abs/1901.07291
10. Levenshtein, V.I.: Binary codes capable of correcting deletions, insertions and reversals. Sov. Phys. Dokl. **10**(8), 707–710 (1966). doklady Akademii Nauk SSSR, vol. 163, no. 4, pp. 845–848 (1965)
11. Lewis, M., et al.: BART: denoising sequence-to-sequence pre-training for natural language generation, translation, and comprehension. CoRR abs/1910.13461 (2019). http://arxiv.org/abs/1910.13461
12. Li, H., Wang, Y., Liu, X., Sheng, Z., Wei, S.: Spelling error correction using a nested RNN model and pseudo training data. CoRR abs/1811.00238 (2018). http://arxiv.org/abs/1811.00238
13. Li, X., Liu, H., Huang, L.: Context-aware stand-alone neural spelling correction. In: Findings of the Association for Computational Linguistics: EMNLP 2020, pp. 407–414. Association for Computational Linguistics, Online (Nov 2020). https://doi.org/10.18653/v1/2020.findings-emnlp.37. https://aclanthology.org/2020.findings-emnlp.37
14. Marcus, M.P., Marcinkiewicz, M.A., Santorini, B.: Building a large annotated corpus of English: the penn treebank. Comput. Linguist. **19**(2), 313–330 (1993)
15. Mays, E., Damerau, F.J., Mercer, R.L.: Context based spelling correction. Inf. Process. Manage. **27**(5), 517–522 (1991). https://doi.org/10.1016/0306-4573(91)90066-U. https://www.sciencedirect.com/science/article/pii/030645739190066U

16. Sakaguchi, K.: robsut-wrod-reocginiton. https://github.com/keisks/robsut-wrod-reocginiton (2017)
17. Sakaguchi, K., Duh, K., Post, M., Durme, B.V.: Robsut wrod reocginiton via semi-character recurrent neural network. In: Proceedings of the Thirty-First AAAI Conference on Artificial Intelligence, 4–9 February 2017, San Francisco, California, USA, pp. 3281–3287. AAAI Press (2017). http://aaai.org/ocs/index.php/AAAI/AAAI17/paper/view/14332
18. Schroff, F., Kalenichenko, D., Philbin, J.: FaceNet: a unified embedding for face recognition and clustering. In: 2015 IEEE Conference on Computer Vision and Pattern Recognition (CVPR). IEEE (2015). https://doi.org/10.1109/cvpr.2015.7298682
19. Vaswani, A., et al.: Attention is all you need. CoRR abs/1706.03762 (2017). http://arxiv.org/abs/1706.03762
20. Wolf, T., et al.: Transformers: state-of-the-art natural language processing. In: Proceedings of the 2020 Conference on Empirical Methods in Natural Language Processing: system Demonstrations, pp. 38–45. Association for Computational Linguistics, Online (Oct 2020). https://doi.org/10.18653/v1/2020.emnlp-demos.6. https://aclanthology.org/2020.emnlp-demos.6
21. Yang, Z., Dai, Z., Yang, Y., Carbonell, J.G., Salakhutdinov, R., Le, Q.V.: Xlnet: generalized autoregressive pretraining for language understanding. CoRR abs/1906.08237 (2019). http://arxiv.org/abs/1906.08237

Assessment of Pharmaceutical Patent Novelty with Siamese Neural Networks

Heba El-Shimy[(✉)], Hind Zantout, and Hani Ragab Hassen

Heriot-Watt University, Dubai, UAE
he12@hw.ac.uk

Abstract. Patents in the pharmaceutical field fulfil an important role as they contain details of the final product that is the culmination of years of research and possibly millions of dollars of investment. It is crucial that both patent producers and consumers are able to assess the novelty of such patents and perform basic processing on them. In this work, we review approaches in the literature in patent analysis and novelty assessment that range from basic digitisation to deep learning-based approaches including natural language processing, image processing and chemical structure extraction. We propose a system that automates the process of patent novelty assessment using Siamese neural networks for similarity detection. Our system showed promising results and has a potential to improve upon the current patent analysis methods, specifically in the pharmaceutical field, by not just focusing on the task from a Natural Language Processing perspective, but also, adding image analysis and adaptations for chemical structure extraction.

Keywords: Document analysis · Siamese neural networks · CNN · LSTM · Optical character recognition · Pharmaceutical patents · Chemical structure extraction

1 Introduction

Innovation in any industrial field is an indicator of economic prosperity [1] especially in developed countries [2]. Innovation stems from Research and Development (R&D) activities in private firms or in the public sector resulting in products that enter the market and generate profit [2]. It was thus necessary to have in place a framework that encourages and incentivises the public and private sectors of any industry to invest in R&D. Intellectual Property (IP) rights and legislation around it were first introduced between 1883 and 1886 to protect "intangible creations" such as those generated by the human mind [3]. Patents are one of many forms of IP, they are granted by a single country to an invention's owner and provide them the right to exclusively use their invention commercially for a limited period of 20 years in most cases, but only in a specific region where the patent was filed for and granted [3]. The patenting system thus guarantees full rights to the inventor, denying any competition the use of

© The Author(s), under exclusive license to Springer Nature Switzerland AG 2023
N. El Gayar et al. (Eds.): ANNPR 2022, LNAI 13739, pp. 140–155, 2023.
https://doi.org/10.1007/978-3-031-20650-4_12

the same product commercially. This advantage is in exchange for the disclosure of the technical details of the invention which helps expand the repository of human knowledge and spurring more research in the same field [4].

For a patent to be granted by the patenting office, it needs to undergo an examination process by an expert in the technical field of the application. The examination process aims to determine the novelty, non-obviousness, and usefulness of an invention [5]. This is usually a rigorous process, requiring considerable time and effort from the examiner especially in assessing an invention's novelty, as this requires searching through a vast database of previous patents in the same domain known as "prior art" [6]. This search process involves around 35,000 to 100,000 patent families and after filtering out non-relevant prior art, up to 5000 patent families remain needing to be personally read by the examiner to assess an invention's novelty [7]. Hence, it is an error-prone process due to many factors, some of which are the large amount of data which may lead to missing important documents during the search, lack of extensive domain knowledge by the examiners creating a backlog of unexamined applications and examiners have to consider ways to speed up the process. [6]. Errors in the examination process may result in granting a patent that conflicts with prior art leading to patent infringement which an act that has legal implications and may lead to the matter taken before a court of law. A settlement can result in compensation being awarded to the rights holder.

Taking the case for pharmaceutical patent similarities, which is the focus of this work, the problem becomes more significant as the pharmaceutical industry is one of the very few fields where the knowledge capture by the patent is often all that is needed to create the product [9]. An important factor to consider is that the cost of inventing a new drug and the R&D costs in the pharmaceutical industry are very high [10]. Most of that cost is in the research phase not in the manufacturing phase as the latter is much cheaper in comparison [11], thus making drug replication by other companies easier and cheaper with little capital investment and that is why there is stringent regulation in place. [9]. Moreover, for a new drug to be approved and released to the market, it needs to undergo years of clinical trials where it cannot be kept a secret, thus patenting is the only means to give the inventor company the ability to make a return on its investments by having the exclusive rights to commercially exploit the product. The previously discussed challenges highlight the importance of carrying rigorous research in prior art databases before investing in the development of a new drug. It also emphasises the role of similarity checks between pharmaceutical patent applications and prior art to avoid infringements and costly settlement fees.

In this paper we propose a novelty assessment system for pharmaceutical patents that is able to determine the degree of similarity between an input patent and prior art. Our contributions could be summarised as follows:

- Developing a pipeline to automate patent document digitization;
- Identifying and extracting images, chemical structures and chemical names and entities from patent text;

– Developing a Siamese deep neural network which combines a Convolutional Neural Network (CNN) with a Bi-directional Long Short Term Memory (Bi-LSTM) Network to detects the degree of similarity between patents.

The rest of this paper is structured as follows: In Sect. 2, a background of methods employed is provided. In Sect. 3, we present our proposed method followed by the achieved results in Sect. 4. Finally, in Sect. 5 we conclude the paper and discuss future work and improvements.

2   Related Work

Patent search and analysis fields have been around for almost two decades and are often referred to in the literature as *patinformatics*. Patinformatics can be defined as "the science of analysing large quantities of patents in order to spot trends and find relationships that would otherwise be difficult to spot when dealing with patents one-by-one" [12]. According to [13], this new discipline involves several tasks and in this section we will briefly discuss relevant work covering the following tasks: patent content analysis, patent relationships and non-textual analysis of documents.

2.1   Patent Content Analysis

A common approach for analysing patent contents is using Subject-Action-Object (SAO)-based analysis. A system proposed by [14] could identify the novelty of a patent given extracted SAO structures and calculates a novelty score for that patent. The authors used a specific software (Knowledgist[TM] by Invention Machine[1]) to extract the semantic structures from the patent documents. They added syntactic data using Part-of-Speech tagging (POS). Then, they calculated similarity measures for the extracted structures; patents with identical structures are linked representing an overlap between them. For calculating the novelty score of a patent, links to antecedent patents are identified; fewer links will result in a higher score. The authors compared their approach to an established one of using citations to assess patent novelty. They found that their proposed approach of using semantic analysis mostly outperforms other methods [14]. SAO structures were also used in work done by [15] where they proposed a system and named it "TechPerceptor". Their method used extracted structures to construct patent maps and networks using patent similarities. The SAO module works in the same way as in [14], and SAO structures are extracted using the same software. The system proposed by [15] allowed the extraction of SAO structures either from the patents abstracts, descriptions or claims, according to the user's preference, as it was noticeable that these sections are the ones rich with detailed information about a patent. They calculated the similarity between pairs of SAO structures, and the similarity score was used to detect patent infringement. If the distance between two patents in the same technology

---

[1] See http://invention-machine.com/ for more information.

field is minimal and, different assignees filed the patents, then this indicates the possibility of infringement and probable conflict.

2.2   Patent Relationships

An established method to measure relationships between patents is citation analysis [16,17]. Each patent has two types of citations; forward citations which means a patent has been cited by other patents that have value in the market, and backward citations which are other antecedent prior art that are referred to by the existing patent [18]. The authors in [18] proposed a model to predict a patent's value based on its forward citations. Their model was fed vector representations of documents as TF-IDF and used Latent Semantic Index (LSI) and Latent Dirichlet Allocation (LDA) to transform the high dimension product of TF-IDF into topic representations and concatenate the results. Then, they measured the distance between different patent representations using the Euclidean measure. Their approach has been successful in patent valuation according to the results of their experiments. Other work by [19] also used citation counts to determine patent quality and value. An Attribute Network Embedding Framework (ANE) was used for creating the citation network and incorporated that with textual analysis using an attention-based CNN. Their results were promising in predicting a patent's quality when tested on the USPTO dataset which is a repository of patents provided by the United Stated Patent and Trademark Office.

Cosine similarity appears extensively in research, for example in [20] the authors chose it solely to evaluate different document feature representation methods. Both [21] and [22] used Cosine similarity along with Euclidean distance and Manhattan distance measures; however, [21] reported that Manhattan distance outperformed other measures. Additionally, both [21] and [22] explored Siamese neural networks for document comparison, [21] opted for DNNs in the form of CNN+LSTM in a Siamese fashion so that one DNN analyses the patent in question and the other analyses other documents in the database. [22] experimented with Siamese shallow CNNs, however, they reported a worse performance compared to other DNNs, but that the trade-off in saving computational costs were noticeable. Authors in [23] expanded upon cosine similarity and put forward a new Textual Spatial Cosine Similarity (TSS) system that does not depend on Natural Language Processing (NLP) techniques for semantic analysis of text. Their proposed system added a spatial dimension to the regular cosine similarity measure where scores of zero were stored for terms that appear in the exact position in sentences and one otherwise. This new method outperformed the regular cosine similarity measure approach, especially in paraphrasing and plagiarism detection systems. In [24], the authors presented a new distance measurement system for detecting a patent's originality, calling it V-score. Their approach measures the technology diversity of a patent through a weighting scheme for the frequency of appearance of classification codes in a given patent. The authors' proposed method gives a higher score if the patent's citation includes more technical classification codes that are non-core which means that the innovation is significant and vice versa for lower scores. They conducted experiments that revealed that their approach is better able to determine a patent's originality and novelty than other citation count-based methods.

2.3   Non-textual Analysis

Semantic analysis of patents may not be sufficient in cases where domain specific structures and figures are included in the patent as is the case in pharmaceutical patents. These rely heavily not only on the text which includes medical terms and drug names but also on graphical representations of chemical compounds. Pharmaceutical patents can also contain images of molecular structures that carry the technical details of a new drug and its active ingredient. So, there is a clear need for a patent analysis method that can effectively extract and understand these non-textual components. Authors in [25] evaluated different chemical Name Entity Recognition (NER) tools. Their chosen tools were the tmChem [26] and ChemSpot [27] which are considered state-of-the-art tools for extracting chemical entities from scientific papers. They assessed these tools without retraining on the four gold standard patent corpora they used; CEMP_T and CEMP_D as cited in [28], chapati[2], and BioS as cited in [29]. Their findings showed that for extracting chemical entities from full text or abstracts, for instance, these tools need to be trained on a similar type of text. Other work by [30] matched search queries to patents in a database through extracting chemical structures from the patent documents and feeding them into another software "name=struct" as cited in [31] to convert these structures into InChI codes. InChI is a standard way to represent chemical substances in the form of text; hence, InChI codes are descriptive for chemical molecules where each molecule has a unique code. They created vector representations for the converted InChI codes so that each code was represented as three vectors, one representing each layer in the compound. Then, they indexed vector representations using the Locality Sensitive Hashing (LSH) algorithm as cited in [32]. Finally, they processed the search query similarly and applied a distance measure to each of the vector representations for the compound in the search query, and the compounds in the database. They used Tanimoto Coefficient as cited in [33] as the distance measure due to it is efficiency in calculating intermolecular similarities. The top matched compounds with the least distance are listed along with the patent documents they are contained in. Similar research by [34] used the International Union of Pure and Applied Chemistry (IUPAC) standard for chemical name identifiers. They used a Conditional Random Field (CRF) model to find chemical entities and calculate the probability of a label sequence given an input sequence, which in this case is the tokenised IUPAC structure of the chemical compound. To convert the found IUPAC structures into names, the authors used OPSIN[3], an open source software. They used a dataset obtained from MEDLINE abstracts for training and testing, as well as some patents for testing. Their experiments showed that their

---

[2] http://chebi.cvs.sourceforge.net/viewvc/chebi/chapati/.

[3] Version of October 11, 2006, http://oscar3-chem.sourceforge.net.

approach was able to recognise chemical entities with an F1 score of 86.5% on a test set constructed from MEDLINE. A different approach was developed by [35] where the authors segmented a page to identify images that contain molecular structures. Following this they used a CNN for the molecule recognition task. Several image datasets were used; the first contained 57 million molecule subsets from the PubChem database [36] paired with the molecule InChI codes, which were then converted to another chemical structure naming standard, Simplified Molecular-Input Line-Entry System (SMILES). A second dataset contained 10 million molecule images rendered using Indigo software [37]. The final dataset contained 1.7 million images of molecules paired with their names curated from the USPTO repository. The authors achieved very good results of 82% molecule recognition accuracy on the Indigo dataset and 77% on the USPTO dataset. They carried out experiments on other datasets, some of which were proprietary, and they achieved good results on those as well.

3   Proposed Method

In this section, we will discuss our proposed approach for an end-to-end patent novelty assessment system that could take as input a patent application and determine its similarity with prior art.

3.1   Data

Data used in this project were obtained from the USPTO [38] open data repository. It contains millions of patent documents since 1790 in PDF format and since 1976 as full text. The website allows access to bulk data products which are curated datasets containing documents filed during certain intervals of time in a structured format such as XML or CSV files. Additionally, the website allows for querying the databases using patent number, title, technology field, applicant(s), and many other criteria, and the query results can be downloaded as individual documents in PDF format. For this work we created our own dataset over two steps:

a) Acquiring full text data files: we downloaded bulk data within a specified time interval. Then, we queried the USPTO database for pharmaceutical patents using certain codes for this technical field, namely, the CPC code. Informed by the literature, pharmaceutical patents have specific CPC codes that start with A61K and that have multiple sub-classes with the code formats ranging from A61K9/00 to A61K48/00. The results of querying the database using these CPC codes were lists of patent applications with their numbers and titles sorted chronologically. Application numbers in the lists were used as a search term on the bulk-downloaded patents so that matching documents were verified and kept while other non-matching documents (non-pharmaceutical) were discarded. The result of this data collection step was 782 pharmaceutical patent applications of a period of three months in XML format.

b) Acquiring original patent files in PDF for extracting non-textual information: the USPTO website search engine was used to query for patents and their matching applications as well as applications cited within those patents. Some other non-matching applications and patents were also downloaded. The goal was to create a dataset containing pairs of documents that are typical (an application and its granted patent), documents that may carry some similarity (a patent and applications cited within) and documents that carry no similarity (an application and a patent on a different pharmaceutical topic). This was a manual process that started with querying the database for patents using their CPC code, downloading their PDF file and using information in the patent document to get its application file number, finally querying the database again to get that application document and download it. The same process was done for applications cited within a certain patent document. The results of this data collection step was 75 pairs of documents (150 documents) labelled as follows; 25 highly similar pairs, 25 cited pairs and 25 dissimilar pairs all in PDF format.

The intuition behind using pairs of documents consisting of a patent in its final form and its application is to create a scenario similar to the workflow of a patent officer where they receive an application and decide if it carries any similarity to prior art (finalised and archived patents). Additionally, this scenario is similar to the process carried out by an inventor where they try to compare their work in—its unfinalised form—with prior art to rule out any possibilities of infringement due to similarity. An improvement in the data collection step could have been using a curated dataset of verified similar and dissimilar patents, but such dataset was not accessible at the time of this work.

3.2   Pipeline for Patent Document Processing

The data collected required more processing to extract useful information. Hence, we created a pipeline that accepts a PDF file as input and outputs a structured, tabular format undergoing the following steps:

a) *Converting PDFs to Images.* In order to perform Optical Character Recognition (OCR)on documents using an open-source library like Tesseract[4], the documents had to be converted to images first. For this task, a python open-source library was used, pdf2image[5], and the conversion to images was done at 600 dpi to ensure highest image quality. Images created from this step were organised into folders named by the patent or application number with a prefix of P- or A- to identify patents from applications. Each patent or application images folder was nested into deeper level of folders for similar, cited or dissimilar pairs of patents. This naming scheme was kept consistent across all following steps and enabled the creation of the final structured dataset.

---

[4] http://code.google.com/p/tesseract-ocr.
[5] https://github.com/Belval/pdf2image.

b) Image Preprocessing. To increase the accuracy of the OCR process, we increased image contrast and cropped pages to exclude the headers and footers to remove any textual noise. The results from this step were saved in a separate folder.

c) Text Extraction. We used a python wrapper for Tesseract's library to extract text from the saved images of patent pages. The settings for this step were: using Tesseract's LSTM option and specifying the English dictionary for text extraction. The results of this step were saved in a separate folder as plain text files.

d) Chemical Structure Extraction. Using an open source library, OSRA[6], chemical structures were extracted and saved from the patent pages image files. The library could also identify the chemical names in their SMILES notation for each extracted structure. The chemical names extracted from these documents were saved separately in plain text files.

e) Chemical Named Entity Recognition and Extraction. ChemSpot[7] is a software developed to recognise and extract chemical named entities from text. It was used in this project to extract chemical names, convert them to their InChI representation and store them in text files to be used for similarity comparisons.

f) Training Dataset Creation. At the end of the patent processing pipeline, two datasets were created. Dataset 1 was created from the XML files containing the full text of patent applications. We traversed the XML nodes which represented different sections in the application document and extracted the title, abstract, summary, detailed description and claims fields. These fields were merged to create one block of text for each document and then saved to a dataframe along with the relevant application number which was saved as a CSV file. Dataset 2 was created from pairs of patent-application PDF files. Using the previously described data enrichment techniques, the final form of the dataset contained the following fields: patent number, matching application number, patent text, application text, patent chemical compounds in SMILES format, application chemical compounds in SMILES format, patent chemical compounds in InChI format, application chemical compounds in InChI format and a label representing similarity. The labels used were: 1—similar, for documents used for applying to the matching patent; 2—cited, for documents referenced in the body of a patent; and 0—dissimilar. Dataset 2 had 75 rows with balanced labels. A variant of this dataset was created during model training experiments with only two labels, where it contained only 50 rows with 25 rows of label '1' for similar documents and 25 rows of label '0' for dissimilar ones. The intuition for creating Dataset 2 is evaluating the models performance on a relatively easier task

---

[6] https://cactus.nci.nih.gov/osra/.

[7] https://www.informatik.hu-berlin.de/de/forschung/gebiete/wbi/resources/chemspot.

with documents being either similar or dissimilar and where we expect to find contrast in similarity score and better separability.

3.3 Creating Word Embeddings

In order to prepare the textual contents of the dataset before being fed to the model, we need to vectorize them to create word embeddings. First, we removed the common English stop words, converted all letters to lower-case and split the text into smaller chunks, usually sentences, that are then referred to as tokens. Then, we tried two approaches for creating word embedding. The first approach was training a model from scratch for any newly introduced patent or application. Doc2Vec discussed in [39] was the tool used to convert tokenised text representing a full patent application into word embeddings. This approach was chosen as pharmaceutical patents contain domain-specific words in addition to legal terms that needed a model trained on such vocabulary. The training carried out in this step created a vector of 500 features for each document. The second approach was using existing word embeddings relevant to the tasks. The US National Center for Biotechnology Information (NCBI) research lab has NLP tools specific to the biomedical field; two tools of interest were pre-trained word-embeddings named BioWordVec and BioSentVec [40] which were available for download. We tried using the latter, BioSentVec, for patent/application text vectorisation in this project as it was more suitable to the lengthy form of the documents. These models were trained on millions of documents in the PubMed and the MIMIC-III Clinical [41] Databases and produced vectors of 700 features for each document.

3.4 Siamese Deep Neural Network Model

For comparing pairs of documents, we used a Siamese neural network architecture, also known as, a twin neural network. The main feature of this architecture is having two identical branches in the network that take two inputs and learn their hidden representation vectors in parallel. Both branches update their weights and parameters at the same time and compare their learnt vectors at the end using a cosine distance function. The output of the network could be interpreted as the semantic similarity of the two input documents [42].

We constructed a Siamese neural network that accepts two documents, a patent and its application in the form of embeddings, as inputs and both vector representations propagate through the model's layers in parallel to extract their features and predict their similarity as output using the cosine distance function. The building blocks for the Siamese Network's branches were a combination of a CNN and a Bi-directional LSTM. Each of the two branches accepts an input vector, then six layers of convolutions and max-pooling with a dropout after each layer. The resulting tensors are passed to a Bidirectional LSTM layer followed by two dense layers and finally a lambda layer to perform cosine distance operation on the resulting tensors and produce a prediction of the similarity between the input documents. Figure 1 depicts the workflow of our proposed diagram and Fig. 2 outlines the architecture of the Siamese neural network used in our work.

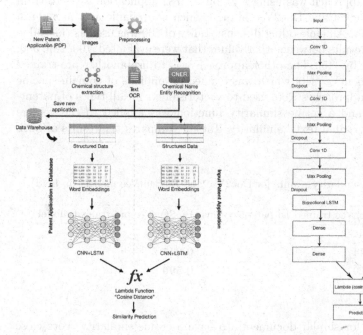

Fig. 1. Diagram of our proposed system.   **Fig. 2.** Structure of siamese neural network

4   Results

4.1   Experimental Setup

We used Python for writing our pipeline code and used the Keras library for writing the code for the Siamese neural network. We automated the pipeline and training using Apache Airflow[8]. We used RMSProp as an optimizer and a learning rate of 0.0001, batch size of 12, dropout of 0.25 and training for 100 epochs. We split our dataset to have 48 sample documents for training, 12 for validation and 15 for testing. We had three classes on the output, namely, similar (1), cited (2) and dissimilar (0). We used contrastive loss as our objective function as in [44].

---

[8] https://airflow.apache.org/.

4.2 Sentence and Document Embeddings Evaluation

As discussed in the previous section, two approaches were used to create word embeddings. One approach was using a corpus of 782 application texts to train a model from scratch using Doc2Vec library which is available as a module in the Gensim[9] library. An embedding dimensionality of 500 was used as this value achieved the best results between other values that were evaluated [100, 200, 300, 500], as suggested by [45]. The other approach was to use another pre-trained model named BioSentVec [40] which was trained on millions of domain-specific documents. Both approaches were used to vectorise text for all types of patent-application pairs and a cosine similarity function was applied to the vectors to determine the original texts' similarity. Table 1 compares the results of both models.

Table 1. Cosine similarity results for Doc2Vec and BioSentVec models (in bold).

Pairs	Doc2Vec trained on patent corpus	BioSentVec pre-trained model
Similar	−0.059	**0.997**
Cited	0.014	**0.947**
Dissimilar	0.020	**0.596**

It is expected for similar documents to have a cosine similarity score close to 1 and for dissimilar ones to be close to zero. The results for BioSentVec were consistent with the expectation, while for Doc2Vec the results were ambiguous. It is thought that one reason Doc2Vec did not achieve good results compared to BioSentVec is due to the large difference in the number of documents used for training the latter (782 vs. 28,714,373). However, BioSentVec model has been trained on biomedical transcripts, thus, we suggest that a Doc2Vec model that is trained on a similar amount of patents may achieve enhanced results as it can grasp chemical, biomedical and legal vocabulary. In order to achieve the best results for the end goal of this work, the pre-trained BioSentVec model has been chosen to create text vectors that were used as input to train the LSTM/CNN models.

4.3 Similarity Detection Model Evaluation

We tracked the training and validation loss and accuracy. Our model reached a training loss of 5.5879×10^{-09}, validation loss of 0.08 and a training and validation accuracy of 100%. We observed that the model converged after around 10 epochs. Evaluating the model on the test split achieved 72% accuracy and −0.06 loss. Figure 3 provides the metrics we tracked while training and the model performance.

---

[9] https://radimrehurek.com/gensim/index.html

4.4   Ablation Studies

We carried out some experiments with only Bi-LSTM in our Siamese model architecture, i.e., removing convolutional layers that preceded the Bi-LSTM layers. We observed in this setting that the model achieved a result of 0.48 for the training loss, 100% training accuracy, 0.375 for the test loss, and a test accuracy of 72%. However, we noticed that the model took longer to converge, about thirty epochs, which is almost triple the time it took our proposed model. This

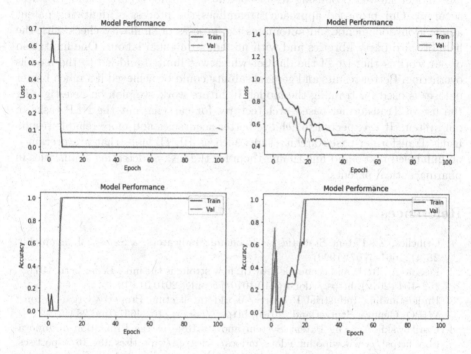

Fig. 3. Performance of our proposed model CNN+Bi-LSTM during training (left) compared to Bi-LSTM-only (right)

could be due to the larger number of trainable parameters in the Bi-LSTM which was $862,256$ versus $377,890$ for the CNN+Bi-LSTM architecture.

Table 2. Summary of our experiments compared to proposed model.

Model	Parameters	Labels	Loss Function	Val. Loss	Acc.	Test Acc.
BiLSTM	$862,256$	3	Contrastive	0.375	100%	72%
CNN+BiLSTM	$377,890$	3	Contrastive	**0.08**	**100%**	**72%**

5   Discussion and Conclusion

In this work, we proposed an end-to-end system for patent novelty assessment that can check patent applications for similarity and possible infringement. The data were collected from the USPTO and underwent several steps of preprocessing in an automated pipeline. We developed a Siamese neural network with CNN and Bi-directional LSTM as the building blocks and trained the network on pairs of patent and their applications that are similar, cited or dissimilar. Our model showed promising results of 100% training accuracy and 72% test accuracy. Our proposed approach streamlines the process of digitizing patent application documents and automates the process of similarity check with the help of third-party libraries and with no human manual labour. One limitation of our work is the size of the dataset which was limited and led to the models overfitting. Better results and generalisability could be achieved if a much bigger dataset is used for training the model. In future work, we plan on investigating the use of Transformer-based architectures for carrying out the NLP tasks in our system. It is expected to achieve better accuracies and more reliable results using Transformer-based architectures such as BERT [46]—that are pretained on unlabeled text—and fine-tuning them for the task of detecting similarities in pharmaceutical patents.

References

1. Griliches, Z.: Patent Statistics as Economic Indicators: a Survey. J. Econ. Lit. **28**(4), 1661–1707 (1990)
2. Pessoa, A.: R&D and economic growth: how strong is the link? Econ. Lett. **107**(2), 152–154 (2010). https://doi.org/10.1016/j.econlet.2010.01.010
3. Understanding Industrial Property. World Intellectual Property Organization - WIPO, Geneva, Switzerland (2016). https://doi.org/10.34667/tind.28945
4. Czajkowski, A.: The Patent System and its Role in the Promotion of Innovation. https://www.wipo.int/edocs/mdocs/africa/en/wipo_tiscs_abv_16/wipo_tiscs_abv_16_t_3.pdf. Accessed 31 Aug 2022
5. Moschini, G.: The economics of traceability: an overview. JRC workshop, Ispra, Italy (2007)
6. Correa, C.M.: Patent examination and legal fictions: how rights are created on feet of clay. In: Drahos, P., Ghidini, G., Ullrich, H. (eds.) Kritika: essays on Intellectual Property. Cheltenham, UK: Edward Elgar Publishing (2015). https://doi.org/10.4337/9781784712068.00010
7. Liu, S.-H., Liao, H.-L., Pi, S.-M., Hu, J.-W.: Development of a patent retrieval and analysis platform - a hybrid approach. Expert Syst. Appl. **38**(6), 7864–7868 (2011). https://doi.org/10.1016/j.eswa.2010.12.114
8. Chaves, G.C., Oliveira, M.A., Hasenclever, L., De Melo, L.M.: Evolution of the international intellectual property rights system: patent protection for the pharmaceutical industry and access to medicines. Cad. Saude Publica **23**(2), 257–267 (2007). https://doi.org/10.1590/S0102-311X2007000200002
9. Lehman, B.: The pharmaceutical industry and the patent system. International Intellectual Property Institute (2003). https://users.wfu.edu/mcfallta/DIR0/pharma_patents.pdf. Accessed 31 Aug 2022

10. DiMasi, J.A., Grabowski, H.G., Hansen, R.W.: Innovation in the pharmaceutical industry: new estimates of R&D costs. J. Health Econ. **47**, 20–33 (2016). https://doi.org/10.1016/j.jhealeco.2016.01.012

11. Kumazawa, R.: Patenting in the Pharmaceutical Industry. In: Prabu, S.L., Suriyaprakasha, T.N.K. (eds.) Intellectual Property Rights. IntechOpen (2017). https://doi.org/10.5772/68102

12. Trippe, A.J.: Patinformatics: tasks to tools. World Patent Inf. **25**(3), 211–221 (2003). https://doi.org/10.1016/S0172-2190(03)00079-6

13. Tseng, Y.-H., Lin, C.-J., Lin, Y.-I.: Text mining techniques for patent analysis. Inf. Process. Manage. **43**(5), 1216–1247 (2007). https://doi.org/10.1016/j.ipm.2006.11.011

14. Gerken, J.M., Moehrle, M.G.: A new instrument for technology monitoring: novelty in patents measured by semantic patent analysis. Scientometrics **91**(3), 645–670 (2012). https://doi.org/10.1007/s11192-012-0635-7

15. Park, H., Kim, K., Choi, S., Yoon, J.: A patent intelligence system for strategic technology planning. Expert Syst. Appl. **40**(7), 2373–2390 (2013). https://doi.org/10.1016/j.eswa.2012.10.073

16. Xia, B., Li, B., Lv, X.: Research on patent document classification based on deep learning. In: 2nd International Conference on Artificial Intelligence and Industrial Engineering - AIIE 2016, AIIE 2016, pp. 308–311. Atlantis Press (2016). https://doi.org/10.2991/aiie-16.2016.71

17. Wanner, L., et al.: Towards content-oriented patent document processing. World Patent Inf. **30**, 21–33 (2008). https://doi.org/10.1016/j.wpi.2007.03.008

18. Liu, X., Yan, J., Xiao, S., Wang, X., Zha, H., Chu, S.M.: On predictive patent valuation: forecasting patent citations and their types. In: 31st AAAI Conference on Artificial Intelligence - AAAI 2017. AAAI 2017, pp. 1438–1444. AAAI press (2017)

19. Lin, H., Wang, H., Du, D., Wu, H., Chang, B., Chen, E.: Patent quality valuation with deep learning models. In: Pei, J., Manolopoulos, Y., Sadiq, S., Li, J. (eds.) DASFAA 2018. LNCS, vol. 10828, pp. 474–490. Springer, Cham (2018). https://doi.org/10.1007/978-3-319-91458-9_29

20. Helmers, L., Horn, F., Biegler, F., Oppermann, T., Müller, K.-R.: Automating the search for a patent's prior art with a full text similarity search. PLoS ONE **14**(3), 1–17 (2019). https://doi.org/10.1371/journal.pone.0212103

21. Pontes, E.L., Huet, S., Linhares, A.C., Torres-Moreno, J.-M.: Predicting the semantic textual similarity with Siamese CNN and LSTM. In: Actes de la Conférence - TALN. vol. 1 - Articles longs, articles courts de TALN, pp. 311–320. ATALA (2018)

22. Yao, H., Liu, H., Zhang, P.: A novel sentence similarity model with word embedding based on convolutional neural network. Concurrency Comput. Pract. Exper. **30**(23) (2018). https://doi.org/10.1002/cpe.4415

23. Crocetti, G.: Textual Spatial Cosine Similarity. arXiv abs/1505.03934 (2015)

24. Harrigan, K.R., Di Guardo, M.C., Marku, E., Velez, B.N.: Using a distance measure to operationalise patent originality. Technol. Anal. Strat. Manage. **29**(9), 988–1001 (2017). https://doi.org/10.1080/09537325.2016.1260106

25. Habibi, M., Wiegandt, D.L., Schmedding, F., Leser, U.: Recognizing chemicals in patents: a comparative analysis. J. Cheminform. **8**, 59 (2016). https://doi.org/10.1186/s13321-016-0172-0

26. Leaman, R., Wei, C.-H., Lu, Z.: a high performance approach for chemical named entity recognition and normalization. J Cheminform. **7**(Suppl 1), S3 (2015). https://doi.org/10.1186/1758-2946-7-S1-S3

27. Rocktäschel, T., Weidlich, M., Leser, U.: ChemSpot: a hybrid system for chemical named entity recognition. Bioinformatics **28**(12), 1633–1640 (2012). https://doi.org/10.1093/bioinformatics/bts183

28. Krallinger, M., et al.: Overview of the CHEMDNER patents task. In: 5th BioCreative Challenge Evaluation Workshop. BC V CHEMDNER Patents Track, pp. 63–75 (2015)

29. Annotated chemical patent corpus: a gold standard for text mining. PLOS ONE **9**(9) (2014). https://doi.org/10.1371/journal.pone.0107477

30. Rhodes, J., Boyer, S., Kreulen, J., Chen, Y., Ordonez, P.: Mining patents using molecular similarity search. In: Pacific Symposium on Biocomputing 2007. PSB 2007, pp. 304–315 (2007)

31. Brecher, J.: Name=struct: a practical approach to the sorry state of real-life chemical nomenclature. J. Chem. Inf. Comput. Sci. **39**(6), 943–950 (1999). https://doi.org/10.1021/ci990062c

32. Indyk, P., Motwani, R.: Approximate nearest neighbors: towards removing the curse of dimensionality. In: 30th Annual ACM Symposium on Theory of Computing, pp. 604–613. Association for Computing Machinery (1998). https://doi.org/10.1145/276698.276876

33. Willett, P., Barnard, J.M., Downs, G.M.: Chemical similarity searching. J. Chem. Inf. Comput. Sci. **38**(6), 983–996 (1998). https://doi.org/10.1021/ci9800211

34. Klinger, R., Kolářik, C., Fluck, J., Hofmann-Apitius, M., Friedrich, C.M.: Detection of IUPAC and IUPAC-like chemical names. Bioinformatics **24**(13), i268–i276 (2008). https://doi.org/10.1093/bioinformatics/btn181

35. Staker, J., Marshall, K., Abel, R., McQuaw, C.M.: Molecular structure extraction from documents using deep learning. J. Chem. Inf. Model. **59**(3), 1017–1029 (2019). https://doi.org/10.1021/acs.jcim.8b00669

36. Kim, S., et al.: Pubchem substance and compound databases. Nucleic Acids Res. **44**(D1), D1202–D1213 (2016). https://doi.org/10.1093/nar/gkv951

37. Indigo Toolkit. Epam Systems. http://lifescience.opensource.epam.com/indigo/. Accessed 31 Aug 2022

38. The United States Patent and Trademark Office. http://patft.uspto.gov/netahtml/PTO/index.html. Accessed 31 Aug 2022

39. Kusner, M.J., Sun, Y., Kolkin, N.I., Weinberger, K.Q.: From word embeddings to document distances. In: 32nd International Conference on Machine Learning - ICML 2015, pp. 957–966. ICML 2015. International Machine Learning Society (IMLS) (2015)

40. Chen, Q., Peng, Y., Lu, Z.: BioSentVec: creating sentence embeddings for biomedical texts. In: 7th IEEE International Conference on Healthcare Informatics - ICHI 2019. ICHI 2019, pp. 1–5. IEEE (2019). https://doi.org/10.1109/ICHI.2019.8904728

41. Johnson, A., Pollard, T., Mark, R.: MIMIC-III Clinical Database (version 1.4). PhysioNet (2016)

42. Chicco, D.: Siamese neural networks: an overview. In: Cartwright, H. (eds) Artificial Neural Networks. Methods in Molecular Biology, 2190. Humana, New York, NY (2021). https://doi.org/10.1007/978-1-0716-0826-5_3

43. Goldberger, A., et al.: Physiobank, physiotoolkit, and physionet: components of a new research resource for complex physiologic signals. Circulation **101**(23), e215–e220 (2000)

44. Hadsell, R., Chopra, S., LeCun, Y.: Dimensionality reduction by learning an invariant mapping. In: 2006 IEEE Computer Society Conference on Computer Vision and Pattern Recognition - CVPR2006, pp. 1735–1742 (2006). CVPR (2006). https://doi.org/10.1109/CVPR.2006.100

45. Yin, Z., Shen, Y.: On the dimensionality of word embedding. In: 32nd Conference on Neural Information Processing Systems - NeurIPS 2018, pp. 887–898 (2018). In: NeurIPS 2018

46. Devlin, J., Chang, M.-W., Lee, K., Toutanova, K. BERT: Pre-training of deep bidirectional transformers for language understanding. In: 2019 Conference of the North American Chapter of the Association for Computational Linguistics: Human Language Technologies - NAACL HLT 2019. NAACL HLT 2019. vol. 1 (Long and Short Papers), pp. 4171–4186. Association for Computational Linguistics (2019). https://doi.org/10.18653/v1/N19-1423

White Blood Cell Classification of Porcine Blood Smear Images

Jemima Loise U. Novia[1], Jacqueline Rose T. Alipo-on[1],
Francesca Isabelle F. Escobar[1], Myles Joshua T. Tan[1,2],
Hezerul Abdul Karim[3], and Nouar AlDahoul[3]

[1] Natural Sciences, University of St. La Salle, Bacolod, Philippines
s1822133@usis.edu.ph
[2] Chemical Engineering, University of St. La Salle, Bacolod, Philippines
mj.tan@usls.edu.ph
[3] Faculty of Engineering, Multimedia University, Cyberjaya, Malaysia
hezerul@mmu.edu.my, nouar.aldahoul@live.iium.edu.my

Abstract. Differentiating white blood cells has been a fundamental part of medical diagnosis as it allows the assessment of the state of health of various organ systems in an animal. However, the examination of blood smears is time-consuming and is dependent on the level of the health professional's expertise. With this, automated computer-based systems have been developed to reduce the time taken for examination and to reduce human error. In this work, an image processing technique was explored to investigate the classification of white blood cells. Through this technique, color and shape features were gathered from segmented nuclei and cytoplasms. Various deep learning algorithms where transfer learning methods were also employed for comparison. Experimental results showed that handcrafted features via image processing are better than features extracted from pre-trained CNNs, achieving an accuracy of 91% when using a non-linear SVM classifier. However overall, deep neural networks were superior in WBC classification as the fine-tuned DenseNet-169 model was found to have the highest accuracy of 93% against all used methods.

Keywords: Classification · CNN · Deep learning · Image processing · Transfer learning · White blood cells

1 Introduction

Blood analysis is a routine laboratory test in veterinary health evaluations [1] which aids in wellness monitoring, disease assessment, and prognostication [2–4]. Aside from automated blood analysis, manual blood smear examination is

The publication of this article was financially supported by Multimedia University, Malaysia. J. Loise U. Novia, J. R. T. Alipo-on, and F. I. F. Escobar contributed equally and share first authorship.

performed for assessment of various blood components. Information such as cell morphology, differential count, and presence of blood parasites can be obtained. The white blood cell (WBC) differential in particular bears great potential clinical value [5], providing information on the wellness of the immune system. However, a limitation of manual analysis is its susceptibility to errors due to its time and work demands. To solve this problem, it is necessary to develop and employ a system for the automated classification of these cells using computer vision technology.

This is made possible by applying computer-aided algorithms, such as image processing and deep learning, to classify WBC at high processing speeds and to yield accurate and precise results.

Image processing is the analysis of digital image data through the application of different mathematical operations using computerized algorithms [6]. It can be used to classify white blood cells based on qualitative characteristics or cell morphology. Here, hand-crafted features are typically generated with pre-processing and segmentation, and are used for classification with a machine learning algorithm. Related works that use traditional algorithms to classify blood cells often use threshold-based methods, such as Otsu's thresholding, in segmenting WBC nuclei [7–10]. Prinyakupt and Pluempitiwiriyawej [11] also proposed a threshold-based method for segmenting leukocyte nuclei and cytoplasms based on morphological properties and cell size which obtained an accuracy of over 90% in two of their datasets. However, the segmentation of WBC cytoplasm is one of the challenging tasks in feature extraction [12–14]. Some techniques used to overcome this were K-means clustering and region growing [15]; K-means clustering and modified watershed algorithm [16]; and principal component analysis (PCA), Otsu's thresholding and area filtering [17].

Deep learning is a subset of machine learning (ML) that originated from artificial neural networks [18]. There are some works that make use of deep learning such as convolutional neural networks (CNNs) in the task of categorizing WBCs. Acevedo et al. [19] proposed a method where they used a transfer learning (TL) approach using the CNNs VGG-16 and Inceptionv3 that were trained and tested with 17,092 images of normal peripheral blood cells. By fine-tuning, the VGG-16 and Inceptionv3 models were able to garner an accuracy of 96% and 95%, respectively. Sharma et al. [20] adopted DenseNet-121 and trained the model with 12,444 images with each class consisting of more than 3000 images. The model yielded a 98.84% accuracy. Banik et al. [21] also proposed a complicated fused CNN architecture that was trained from scratch on a dataset consisting of 10,253 augmented WBC images and it was able to achieve an accuracy of 98.61%. Lin et al. [22] combined a K-means clustering algorithm and a watershed algorithm for WBC extraction for CNN classification. Such method yielded a 96.24% overall accuracy, outperforming SVM and back-propagation neural network approaches cited in the study.

While such models promise state-of-the-art accuracy, building a good model requires a huge dataset and this can be computationally expensive [23]. One approach that aims to address this issue is TL. Since it makes use of pre-trained algorithms on a new task [24], a large dataset and long training time are no longer required to train the model. Some WBC classification problems have been addressed using this method. Bagido et al. [25] performed classification using pre-trained models and they showed promising results where the Inception ResNetV2 obtained the highest accuracy (98.4%). Shahin et al. [26] also proposed a model called WBCsNET which was initially built from scratch and was re-employed as a pre-trained model to perform classification on a limited WBC dataset. After the experiment, the proposed model was able to achieve a 96.1% accuracy. A combination of deep learning and traditional machine learning can be also performed through TL, as shown by Baby et al. [27]. They explored the use of TL where they used a VGG-16 model as feature extractor and the features that were extracted were trained and tested using a K-Nearest Neighbor (KNN) model. The proposed method was able to perform with 82.35% accuracy. Jung et al. [28] compared the classification performance of the softmax classifier of W-Net with SVM. Although W-Net demonstrated superior performance, W-Net-SVM was still able to perform with promising accuracy (95%).

In this study, an image processing technique was explored as a feature extractor in classifying white blood cells. Shape and color features were generated from the segmented nucleus and cytoplasm. State-of-the-art CNN-based pre-trained models were also employed through TL techniques for comparison where the softmax classifier of the CNN-based models were compared with the SVM.

2 Methodology

2.1 Dataset

We prepared the dataset [29] used for this project under the supervision of a licensed veterinarian. Data acquisition involved capturing the digital images of the blood smear under the microscope. This was specifically done with a smartphone mounted on an Olympus® CX23 compound microscope set under oil immersion magnification. Prior to image acquisition, peripheral blood films were prepared with blood samples collected from juvenile pigs and were fixed with Hema-Quick (Pappenheim) staining.

A total of 655 digital images were acquired for processing. Each image was extracted of a number of samples wherein each sample image contained only a single white blood cell. WBCs without overlapping cells were selected with each sample cropped to a uniform size of 722 × 422. From the 655 images, a total of 1526 images were extracted, being composed of an uneven amount of each WBC class. Each image was identified by a licensed medical technologist. The cell types considered for the study were basophil, eosinophil, lymphocyte, monocyte, and neutrophil. Figure 1 displays a set of examples of white blood cells of each class.

In SVM using image processing, images were retained of the size 722 × 422. Whereas for TL with CNNs, the images were further cropped such that only the target WBC comprised most of the image. All were uniformly cropped to a size of 300 × 300. This difference in datasets was performed to consider that CNNs

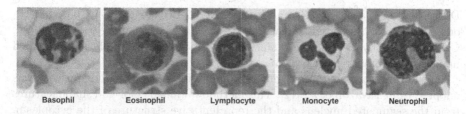

Fig. 1. Example images of white blood cells of each type: basophil, eosinophil, leukocyte, monocyte, and neutrophil

may take into account the background cells as part of the target cell's features.
The used image processing methods directly extracted only specific features,
thus it needed no further processing.

Image augmentation via Keras preprocessing layers was applied to both
datasets to resolve the class imbalance. The augmentation process included random affine transformations consisting of shearing, rotation, and random flip.
After the process, each class was balanced to 500 images, except for the basophil
class which had 467 images. Overall, each dataset consisted of 2467 images for
training and testing the algorithms. Table 1 shows the initial numbers of images
per class, the amount of augmented images generated, and the final number of
images used for the study.

Table 1. Number of images per class and the respective amount of data created for
the study

Class	Original images	Augmented images	Final dataset
Basophil	20	447	467
Eosinophil	82	418	500
Lymphocyte	905	None	500
Monocyte	82	418	500
Neutrophil	319	181	500

70% was allotted randomly for training, and 30% for testing in the classification using image processing and pre-trained CNNs as feature extractors.
For the fine-tuning approach, the dataset was divided into training, validation,
and testing sets where each set was allocated 70%, 15%, and 15% of the data,
respectively.

2.2 Model Implementation

The study adopted two main algorithms in classifying blood cells: image processing and CNN-based methods involving pretrained models VGG-16, VGG-19,

DenseNet-121, DenseNet-169, and EfficientNetv2-B0. For the CNN-based methods, TL approaches including feature extraction and fine-tuning were employed. Comparison between the performance of the softmax classifier of the pre-trained models were compared with SVM.

For image processing, the segmentation and feature extraction approach was adopted from Tavakoli et al. [12]. This method obtained color and shape features from the segmented nucleus and the representative structure of the cytoplasm. Color balancing techniques were applied to the image to ensure that only the target subjects were visible prior to segmentation; this involved measuring the mean of the RGB and grayscale values of the image followed by conversion and manipulation of the CMYK and HLS color spaces. Briefly, The M component, containing a higher intensity for the nucleus of the WBC, was subtracted from the K component which contained a greater intensity for the RBC and cytoplasm. The subtraction generated an image with close to zero pixel values for the nucleus. Meanwhile, the minimum of M component from the CMYK colorspace and S component from the HLS colorspace was computed, producing an image where the values of the intensity of RBC and background are close to zero. The difference from the K component and M component and the result from the M component and S component were then subtracted. Thus, the intensity of other cells, cytoplasm, and the background were eliminated: only the nucleus' structure was visible in the image. For segmentation, Otsu's thresholding was applied. The nucleus was first extracted, from which the convex hull was used to acquire the representative of the cytoplasm (ROC). This was done instead of extracting the whole cytoplasm due to challenges in segmentation accuracy. A diagram summarizing the segmentation portion of the image processing method is seen in Fig. 1. Data from image processing was used for SVM classification. Prior to classification, the features extracted were normalized using min-max normalization. For the SVM, the radial basis function with a regularization parameter C of 26 and a degree of 1 was set as the hyperparameter to train the model properly. The selection of the optimal hyperparameter was derived from GridSearchCV.

For TL, two approaches were employed: feature extraction and fine-tuning. This method involved resizing all input images to $224 \times 224 \times 3$ in accordance with the set value when pre-trained on ImageNet. The Image-Net pre-trained top fully connected layers of the models were omitted, and classification was performed with a nonlinear SVM classifier in the feature extraction approach and with modified fully connected layers in the fine-tuning approach.

VGGNet is a CNN which consists of a Rectified Linear Unit (ReLU) after each layer with 3×3 convolution filters. VGG-16 and VGG-19 vary in the number of convolutional layers, with VGG-16 having 13 layers and the latter with 16.

DenseNet is composed of layers connected directly to every other layer in a feed-forward fashion. Four dense blocks form this model, in which the number of layers within each block varies between DenseNet-121 (6, 12, 24, 16) and DenseNet-169 (6, 12, 32, 32). Each layer corresponds to the composite function operations that include the batch normalization, ReLU, and a convolution.

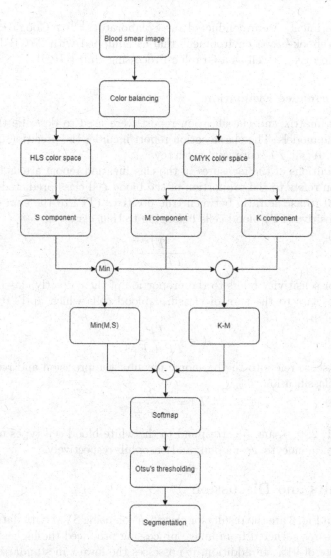

Fig. 2. A block diagram of the segmentation phase of the image processing method

EfficientNetv2-B0 is a variation of EfficientNet with fewer parameters (24M) than its older model (43M). Its last stride-1 stage was excluded to avoid the large parameter size and memory access overhead. The model architecture uses both inverted residual blocks (MBConv) and fused inverted residual blocks (Fused-MBConv) of 3 × 3 kernel sizes but with more layers reaching a total of 42. Images during training are progressively resized into a larger image along with the application of regularization.

All experiments were conducted on Colaboratory™ by Google, an online Jupyter notebook-based environment that is equipped with 25 GB RAM and 150 GB main disk, as well as a Graphic Processing Unit (GPU).

2.3 Performance Evaluation

A confusion matrix and classification report were used to describe the performance of the models. The classification report includes the precision or positive predictivity, recall, F1-score, and accuracy.

The definitions of the measures in the classification report are as follows:

Precision refers to the proportion of the blood cell class predicted correctly, also referred to as the ratio between true positives (TP) to the sum of the TP and false positive (FP) blood cells belonging to that class:

$$P = \frac{TP}{TP + FP} \tag{1}$$

Recall or sensitivity refers to the proportion of the correctly classified blood cells of one class to the sum of classified blood cells which are TP and false negative (FN):

$$R = \frac{TP}{TP + FN} \tag{2}$$

The F1-score refers to the harmonic mean of the precision and recall of the blood cell classification:

$$F1 = \frac{2}{\frac{1}{R} + \frac{1}{P}} \tag{3}$$

Labels 1, 2, 3, 4, and 5 correspond to the white blood cell types neutrophil, lymphocyte, monocyte, eosinophil, and basophil, respectively.

3 Results and Discussion

Presented in Fig. 3 are the results for classification using SVM with data augmentation. Features garnered from image processing produced the highest accuracy and F1-score of 91%. In addition, it possesses the lowest in standard deviation indicating that it offers more precise and consistent performance than CNN-based feature extraction. Even though this approach was more time consuming, the features obtained were more precise and relevant for classification. Contrary to pretrained CNNs, it is also more interpretable. Image processing lends to more control of the specific features for extraction (nucleus, ROC, convex hull, color features) whereas CNNs are dependent on their own capacity.

Amongst the selected CNNs, VGG16, DenseNet169, and EfficientNetvB0 performed best with 90% for both accuracy and F1-score. Overall, CNN-based feature extraction exhibits more stochastic behavior than image processing-SVM with VGG19 and DenseNet169 having a relatively high standard deviation compared to the other models. The reason to this is unclear as the transparency and interpretability are known to be limitations of deep learning models [30].

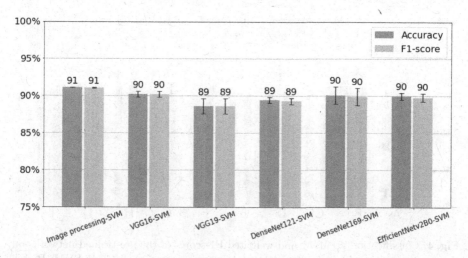

Fig. 3. SVM classification accuracies obtained from feature extraction through image processing and the selected pretrained CNNs after applying data augmentation.

Figure 4 provides a graphical comparison of the effect of data augmentation on TL techniques. It is apparent that the use of data augmentation (DA) benefited the model for all methods. Letters A, C, E, G, and I represent the performance metrics measured when TL CNNs was used for feature extraction followed by SVM classification. Without data augmentation, DenseNet169 with SVM and EfficientNetvB0 with SVM produced the highest accuracy of 83% with the lowest at 80% using VGG19. After data augmentation, there was a significant improvement in performance. VGG16, DenseNet169, and EfficientNetvB0 were found to have the highest accuracy and F1-score when used with SVM classification.

Letters B, D, F, H, and J in Fig. 4 presents the accuracy and weighted F1-score of fine-tuned (FT) CNNs for classification before and after using data augmentation. Altogether, the CNNs performed better when used for classification rather than feature extraction alone. FT DenseNet169 had the highest result of 85% accuracy, followed by FT VGG-19 with 84% when using the original(O) dataset. With data augmentation FT DenseNet169 still show superiority having an accuracy and weighed F1-score of 93%. In fine tuning pre-trained deep CNNs, the performance may be dependent on the architecture of the model itself or the pre-learned weights themselves, thus, models with more depth or more parameterized layers are better compared to those with less depth in terms of generalization of data [31, 32]. Overall, DenseNet169 outperforms all selected CNNs for both feature extraction and classification tasks. Furthermore, it also achieved the highest results for both original and augmented datasets. DenseNet169, having more parameterized layers than the other CNNs used in the study, evidently outperformed other models when fine-tuning was applied.

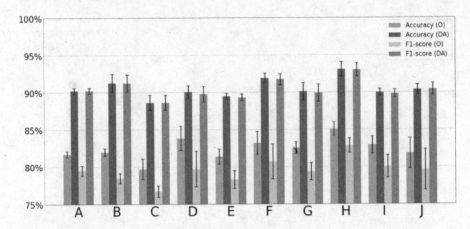

Fig. 4. Classification accuracy and weighted F1-scores of the pre-trained networks on the datasets with(DA) and without(O) data augmentation. **A.** VGG16-SVM, **B.** FT VGG16, **C.** VGG19-SVM, **D.** FT VGG19, **E.** Densenet121-SVM, **F.** FT DenseNet121, **G.** DenseNet169-SVM, **H.** FT DenseNet169, **I.** EfficientNetvB0-SVM, **J.** FT Efficient-Netv2B0

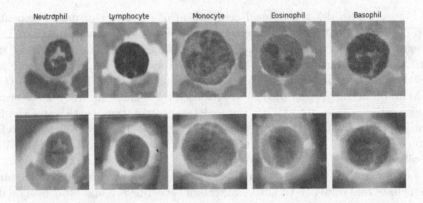

Fig. 5. Grad-CAM of the fine-tuned DenseNet-169.

Figure 5 features the visualization of the attention map output of the fine-tuned DenseNet-169, which highlights the area of highest attention when the network undergoes classification decision with warm color hues. The main focus, overlayed with red, appeared to be the nucleus of the WBC, followed by the cytoplasm.

Table 2 shows the performance obtained by traditional image processing with SVM and the five fine-tuned pre-trained CNNs after data augmentation. The FT DenseNet-169 garnered the highest accuracy (93%). Among the five types of white blood cells, the lowest precision, recall, and F1-scores were for the classification of lymphocytes and monocytes. This may be due to the high variability in the morphology of these cells observed during data acquisition. In addition, structures of lymphocytes and monocytes exhibited circular to irregular blob shapes which would often resemble each other.

Table 2. Best classification results from each of the selected models. Neu = Neutrophil, Lym = Lymphocyte, Mon = Monocyte, Eos = Eosinophil, Bas = Basophil

WBC	Neu			Lym			Mon			Eos			Bas			
Metric	P	R	F1	P	R	F1	P	R	F1	P	R	F1	P	R	F1	Acc
IP-SVM	91	91	91	88	89	88	86	**90**	**88**	93	93	93	**97**	93	94	91
FT VGG-16	90	92	91	87	92	89	89	74	81	93	94	93	97	**98**	**97**	91
FT VGG-19	92	89	90	87	88	87	83	71	76	91	95	93	96	**98**	**97**	90
FT DenseNet-121	92	87	89	87	**93**	**90**	**91**	76	83	94	**98**	96	94	**98**	**97**	92
FT DenseNet-169	**94**	**94**	**93**	**89**	91	**90**	88	83	86	**96**	95	**97**	**97**	**98**	**97**	**93**
FT EfficientNetv2-B0	93	88	91	84	90	87	**90**	78	83	95	95	95	90	97	93	90

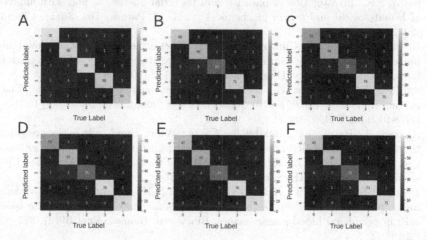

Fig. 6. Classification results of five models presented in Table 2 through a confusion matrix. **A.** traditional image processing, **B.** VGG-16, **C.** VGG = 19, **D.** DenseNet-121, **E.** DenseNet-169, and **F.** EfficientNetv2-B0.

Figure 6 presents the confusion matrices of the different models presented in Table 2.

4 Conclusion

In the comparison of traditional ML and deep learning for porcine WBC classification, deep learning with pre-trained CNNs outperformed all selected approaches. Based on accuracy, TL using fine-tuned DenseNet-169 was superior achieving a score of 93%. Findings also support that when employing traditional ML such as SVM, using data extracted from image processing techniques are more effective than TL techniques. Application of image processing for SVM attained an accuracy of 91%. Overall the study gained promising results for application in providing assistance to cell identification. Further improvement of such methods may aid in progressing veterinary medicine for more efficient systems that may equip facilities for better animal care. It is recommended that the dataset be increased, especially for neutrophils, eosinophils, and basophils.

References

1. Harvey, J.W.: Chapter 1 - introduction to veterinary hematology. In: Harvey, J.W. (ed.) Veterinary Hematology, pp. 1–10. W.B. Saunders, Saint Louis (2012). https://doi.org/10.1016/B978-1-4377-0173-9.00001-4
2. Ježek, J., et al.: The influence of age, farm, and physiological status on pig hematological profiles. J. Swine Health Prod. **26**, 72–78 (2018)
3. Karalyain, Z., et al.: Evidence of hemolysis in pigs infected with highly virulent African swine fever virus. Vet. World **9**(12), 1413–1419 (2016)
4. Kalai, K., Nehete, R.S., Ganguly, S., Ganguli, M., Dhanalakshmi, S., Mukhopadhayay, S.K.: Investigation of parasitic and bacterial diseases in pigs with analysis of hematological and serum biochemical profile. J. Parasit. Dis. **36**(1), 129–134 (2012)
5. Beckman, A.K., et al.: Clinician-ordered peripheral blood smears have low reimbursement and variable clinical value: a three-institution study, with suggestions for operational efficiency. Diagn. Pathol. **15**(1), 112 (2020)
6. da Silva, E., Mendonca, G.: Digital image processing. In: The Electrical Engineering Handbook, pp. 891–910. Academic Press (2005). https://doi.org/10.1016/B978-012170960-0/50064-5
7. Ahasan, R., Ratul, A.U., Bakibillah, A.S.M.: White blood cells nucleus segmentation from microscopic images of strained peripheral blood film during leukemia and normal condition. In: 2016 5th International Conference on Informatics, Electronics and Vision (ICIEV), pp. 361–366 (2016)
8. Hegde, R.B., Prasad, K., Hebbar, H., Singh, B.M.K.: Comparison of traditional image processing and deep learning approaches for classification of white blood cells in peripheral blood smear images. Biocybern. Biomed. Eng. **39**(2), 382–392 (2019)
9. Liu, Z., et al.: Segmentation of white blood cells through nucleus mark watershed operations and mean shift clustering. Sensors (Basel) **15**(9), 22561–22586 (2015)
10. Gautam, A., Bhadauria, H.: Classification of white blood cells based on morphological features. In: 2014 International Conference on Advances in Computing, Communications and Informatics (ICACCI), pp. 2363–2368 (2014). https://doi.org/10.1109/ICACCI.2014.6968362
11. Prinyakupt, J., Pluempitiwiriyawej, C.: Segmentation of white blood cells and comparison of cell morphology by linear and naïve bayes classifiers. BioMed. Eng. OnLine **14**(1) (2015). https://doi.org/10.1186/s12938-015-0037-1
12. Tavakoli, S., Ghaffari, A., Kouzehkanan, Z.M., Hosseini, R.: New segmentation and feature extraction algorithm for classification of white blood cells in peripheral smear images. Sci. Rep. **11**(1), 19428 (2021)
13. Al-Kofahi, Y., Zaltsman, A., Graves, R., Marshall, W., Rusu, M.: A deep learning-based algorithm for 2-D cell segmentation in microscopy images. BMC Bioinform. **19**(1), 365 (2018)
14. Dimopoulos, S., Mayer, C.E., Rudolf, F., Stelling, J.: Accurate cell segmentation in microscopy images using membrane patterns. Bioinformatics **30**(18), 2644–2651 (2014). https://doi.org/10.1093/bioinformatics/btu302
15. Sarrafzadeh, O., Dehnavi, A.: Nucleus and cytoplasm segmentation in microscopic images using K-means clustering and region growing. Adv. Biomed. Res. **4**(1), 174 (2015). https://doi.org/10.4103/2277-9175.163998   .

16. Vard, A., Ghane, N., Talebi, A., Nematollahy, P.: Segmentation of white blood cells from microscopic images using a novel combination of k-means clustering and modified watershed algorithm. J. Med. Sig. Sens. **7**(2), 92 (2017). https://doi.org/10.4103/2228-7477.205503

17. Makem, M., Tiedeu, A.: An efficient algorithm for detection of white blood cell nuclei using adaptive three stage pca-based fusion. Inform. Med. Unlock. **20**, 100416 (2020). https://doi.org/10.1016/j.imu.2020.100416

18. Shrestha, A., Mahmood, A.: Review of deep learning algorithms and architectures. IEEE Access **7**, 53040–53065 (2019). https://doi.org/10.1109/ACCESS.2019.2912200

19. Acevedo, A., Alférez, S., Merino, A., Puigvì, L., Rodellar, J.: Recognition of peripheral blood cell images using convolutional neural networks. Comput. Methods Programs Biomed. **180**, 105020 (2019). https://doi.org/10.1016/j.cmpb.2019.105020

20. Sharma, S., et al.: Deep learning model for the automatic classification of white blood cells. Comput. Intell. Neurosci. **2022**, 1–13 (2022). https://doi.org/10.1155/2022/7384131

21. Banik, P.P., Saha, R., Kim, K.-D.: An automatic nucleus segmentation and CNN model based classification method of white blood cell. Expert Syst. Appl. **149**, 113211 (2020). https://doi.org/10.1016/j.eswa.2020.113211

22. Lin, L., Wang, W., Chen, B.: Leukocyte recognition with convolutional neural network. J. Algorithms Comput. Technol. **13**, 174830181881332 (2018). https://doi.org/10.1177/1748301818813322

23. Sahlol, A.T., Kollmannsberger, P., Ewees, A.A.: Efficient classification of white blood cell leukemia with improved swarm optimization of deep features. Sci. Rep. **10**(1) (2020). https://doi.org/10.1038/s41598-020-59215-9D

24. Zhu, C.: Pretrained language models. In: Machine Reading Comprehension, pp. 113–133 (2021). https://doi.org/10.1016/b978-0-323-90118-5.00006-0

25. Bagido, R.A., Alzahrani, M., Arif, M.: White blood cell types classification using deep learning models. Int. J. Comput. Sci. Netw. Secur. **21**(9), 223–229 (2021). https://doi.org/10.22937/IJCSNS.2021.21.9.30

26. Shahin, A.I., Guo, Y., Amin, K.M., Sharawi, A.A.: White blood cells identification system based on convolutional deep neural learning networks. Comput. Methods Programs Biomed. **168**, 69–80 (2019). https://doi.org/10.1016/j.cmpb.2017.11.015

27. Baby, D., Devaraj, S.J., Raj, M.M.A.: Leukocyte classification based on transfer learning of vgg16 features by k-nearest neighbor classifier. In: 2021 3rd International Conference on Signal Processing and Communication (ICPSC), pp. 252–256 (2021). https://doi.org/10.1109/ICSPC51351.2021.9451707D

28. Jung, C., Abuhamad, M., Mohaisen, D., Han, K., Nyang, D.: Wbc image classification and generative models based on convolutional neural network. BMC Med. Imag. **22**(1) (2022). https://doi.org/10.1186/s12880-022-00818-1

29. Alipo-on, J.R., et al.: Dataset for machine learning-based classification of white blood cells of the Juvenile Visayan Warty Pig. https://doi.org/10.21227/3qsb-d447. https://dx.doi.org/10.21227/3qsb-d447

30. Camilleri, D., Prescott, T.: Analysing the limitations of deep learning for developmental robotics. In: Mangan, M., Cutkosky, M., Mura, A., Verschure, P.F.M.J., Prescott, T., Lepora, N. (eds.) Living Machines 2017. LNCS (LNAI), vol. 10384, pp. 86–94. Springer, Cham (2017). https://doi.org/10.1007/978-3-319-63537-8_8

31. Srivastava, R.K., Greff, K., Schmidhuber, J.: Training very deep networks. In: Proceedings of the 28th International Conference 20n Neural Information Processing Systems - Volume 2. NIPS 2015, pp. 2377–2385. MIT Press, Cambridge, MA, USA (2015)

32. He, K., Zhang, X., Ren, S., Sun, J.: Deep residual learning for image recognition. In: 2016 IEEE Conference on Computer Vision and Pattern Recognition (CVPR), pp. 770–778 (2016). https://doi.org/10.1109/CVPR.2016.90

Medical Deepfake Detection using 3-Dimensional Neural Learning

Misaj Sharafudeen(ID) and S. S. Vinod Chandra(✉)(ID)

Department of Computer Science, University of Kerala, Kerala, India
{misaj,vinod}@keralauniversity.ac.in

Abstract. In recent years, Generative Adversarial Networks (GAN) have underlined the necessity for exercising caution in trusting digital information. Injection and removal of tumorous nodules from medical imaging modalities is one method of maneuvering deepfakes. The inability to acknowledge medical deepfakes can result in a substantial impact on healthcare procedures or even cause of death. With a systematic case study, this work seeks to address the detection of such assaults in lung CT (Computed Tomography) scans generated using CT-GANs. We experiment with machine learning methods and a novel 3-dimensional deep neural architecture on the topic of differentiating between tampered and untampered data. The proposed architecture on the CT-GAN dataset attained a remarkable accuracy of 91.57%, sensitivity of 91.42%, and specificity of 97.20%. Sectioned data cubes containing the affected region of interest seem to perform better compared to raw CT slices with a gain of approximately 20%. Furthermore, 3DCNN outperforms its 2-dimensional counterpart as it extracts temporal features unlike the spatial relationship insufficient for medical data processing. The outcomes of this research reveal that nodule injection and removal manipulations in complicated CT slices may be recognized with a high degree of precision.

Keywords: Deepfake detection · Computed tomography · Fake cancer nodules · Machine learning · Deep learning

1 Introduction

Digital fabrication and multimedia creation or manipulation are not new concepts. Deep Neural Networks (DNNs) have advanced rapidly in recent years, making it easier and faster to produce convincing false data. Deepfakes are artificially synthesized multimedia content generated using highly intelligent networks named Generative Adversarial Networks (GANs) [1]. They produce new samples that imitate an existing data set. Since the first documented example of deepfake videos in the year 2017, there has been a surge of interest in deepfake detection among the research community [2].

A GAN comprises two deep neural networks: a generator network that generates signals of any dimension from random noise, and a discriminator network

© The Author(s), under exclusive license to Springer Nature Switzerland AG 2023
N. El Gayar et al. (Eds.): ANNPR 2022, LNAI 13739, pp. 169–180, 2023.
https://doi.org/10.1007/978-3-031-20650-4_14

that determines if the data given is real or fraudulent. During training, the discriminator is supplied with a mixture of actual data (real) and produced deepfakes (from the generator). Both the generator and the discriminator increase in performance as training progresses until the generator is able to produce realistic fakes. Every year, new and improved modified GANs are released. The underlying principle, though, remains the same.

These generative algorithms mostly focus on spawning fake facial data due to the convincing and credible nature of highly realistic synthesized face data [3]. With the rise in deepfake threats, it is equally important to invest deeply in the detection of artificially synthesized data. Various machine learning and deep learning techniques are still being researched.

The consistency of the biological signals such as the rate of blinking of eyes and lip movements are measured to produce a set of features representing bonafide and spoofed data [4]. Moreover, statistical and handcrafted features extracted in the form of spatial features, textual information and visual artifacts seem to perform better compared to biological signals [5]. Deep learning models intuitively have the ability to recognize specific artifacts generated by a data generation process. In [6], visual and audio features extracted from a deep neural network are contrastively learned for categorizing fake from real. Several advanced models including transformers and attention networks are being widely used lately [7]. However, none have been yet categorized as a breakthrough in the context of a generalized deepfake detection model.

Recently, this technique was employed to produce additional medical data for research purposes [8]. Thambawita et al. proposed the synthesis of complex Electrocardiograms (ECG) using a WaveGAN [9]. Frid-Adar et al. proposed a Deep Convolutional GAN (DCGAN) to synthesize high-definition CT scans of skin lesions [10]. Likewise, several works propose artificially synthesizing skin lesions, brain tumors and difficult-to-reproduce histopathological data [11][12][13]. While synthesized data offers a solution to the medical privacy dilemma by allowing researchers to share data, it is always vulnerable to criminal use.

GANs have lately been utilized to construct medical deepfakes and machine learning-based diagnosis tools to deceive medical practitioners by adding or erasing symptoms and indications of medical conditions [14]. This domain has not received enough investigation and might become a futuristic weapon. Jekyll, a framework proposed by Mangaokar et.al demonstrates a style transfer mechanism for medical image generation [15]. In [16], the content of 3D medical image data is fraudulently altered in a realistic and automated manner. CT (Computerized Tomography) is a medical imaging modality that enables images of bones, and tissues. Mirsky et al. developed CT-GAN, a conditional generative adversarial network (CGAN) capable of adding or removing cancerous nodules from lung CT scans. Three experienced radiologists and an Artificial Intelligence (AI) detection model examined the manipulated scans. Over 90% of the tampered scans were misdiagnosed as real by the doctors, whereas, the detection model failed in identifying any of the fraudulent samples. Solaiyappan et.al investigated several machine learning and pre-trained Convolution Neural Networks (CNNs) on CT-GAN generated data [17]. The lack of data and simplicity of the proposed models had an inverse effect on detection accuracy. Furthermore, considering the task as a multi-class classification revealed the underwhelming precision of the tampered category.

We propose a novel 3-dimensional deep neural architecture for detecting artificially manipulated 3D imaging modality. In addition to the DL model, several machine learning models have also been experimented with raw data for comparison. Furthermore, the compilation and statistics of the medical deepfake dataset are presented.

2   Dataset

We utilized the CT-GAN dataset [16] and the LIDC-IDRI [18] dataset from the Cancer Imaging Archive in this study. CT-GAN dataset consists of tampered (fake) and untampered (real) data representing the False Benign (FB), False Malignant (FM), True Benign (TB), and True Malignant (TM) cases. All radiographs are stored in DICOM format. True Malignant cases are the appearance of real cancerous nodules in CT scans of the lung region, whereas, True Benign regions are the non-cancerous scan regions. The False Benign class constitutes artificially removed nodules from True Malignant cases. Similarly, the False Malignant class comprises artificially synthesized nodules in True Benign regions. Furthermore, LIDC-IDRI consists of only real untampered data in the form of True Benign and True Malignant nodules.

Due to data imbalance and lack of inconsistent data in CT-GAN, more False cases were generated using the augmentation technique. We used image transformations that would maintain the embedded features. Specifically, the rotation, flip, emboss and sharpen spatial transformations ensured that the orientation and appearance of image data were only altered. Transformations such as shear and optical distortion were avoided. To balance the dataset, True cases were included from the LIDC-IDRI repository. The number of total cases belonging to each class and their proportion concerning the entire dataset is shown in Table 1. The newly compiled dataset is made freely available in *http://www.mirworks.in/downloads.php*.

Table 1. Details on the CT-GAN Augmented datasets

Class Type	Label	Total	Percentage
FB- False Benign	0	216	26.15
FM- False Malignant	1	205	24.70
TB- True Benign	2	202	24.45
TM- True Malignant	3	205	24.70

Scan identification numbers (Scan IDs) relating to the patient under-diagnosis are also included in the dataset. Though the ID has no significant relevance, it could be used to uniquely identify each CT scan set. Each scan set may contain multiple nodules (true, removed, or injected) that are represented using the (z, x, y) coordinates denoting the slice number and (x,y) coordinates of the nodule center. Figure 1 exhibits a few samples of the 4 classes (eight slices of each case) and is evidence of the realistic nature of the fake samples (FB and FM).

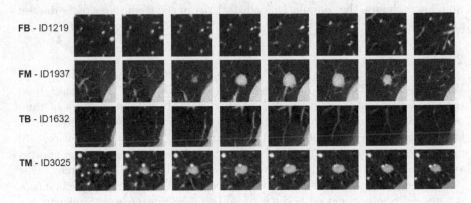

Fig. 1. 8 sliced 64×64 cut cubes of select samples from the CT-GAN augmented dataset

3   Proposed Methodology

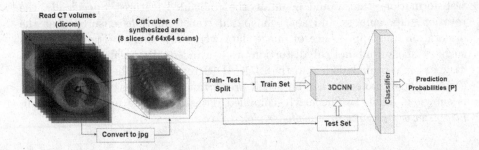

Fig. 2. Framework of the 3D CTGAN detector using 3DCNN

Figure 2 depicts the general outline of the medical deepfake detector. Original CT scans are initially read and converted to *.jpg* format. We used the (z,x,y) coordinates of the affected region to cut cubes of size 8×64×64 (8 slices of 64 × 64 dimensioned scan images). The cubes are then split to train, validation, and test sets in the ratio of 8:1:1. While 8 parts of the entire dataset are absorbed in training, 1 part each validates and tests the trained model on unseen data.

We trained a set of machine learning classifiers and a 3-Dimensional Convolution Neural Network (3DCNN) separately with the processed data cubes.

3.1   Detection Using Machine Learning

Machine Learning (ML) is the subset of AI that uses computational statistics extracted from previously observed data to improve performance on new data. ML analysis attempts to serve as the baseline model for our classification model. The experiment is intended to measure and compare the effect of raw data on ML algorithms as compared to DL models. Each of the data cubes was initially flattened to a 32768 (8 * 64 * 64) vector which is fed into the algorithms as single data points. Linear Support Vector Machine (SVM) for multiple classes is trained and evaluated to find a hyperplane that separates each class member maximally. K Nearest Neighbour (KNN) connects data points that are close together. K is the number of nearby data points with which the data point to be categorized is compared. Additionally, the sample is tagged with the most popular class. Logistic Regression is an approach that uses binary relationships expressed as probability values ranging from 0 to 1. The multiclass problem is approached as a one-versus-many problem, with one class being categorized in relation to the others. We also engaged decision trees and random trees, algorithms in which data is repeatedly and hierarchically categorized depending on a parameter. Decisions taken by decision trees solely depend on the depth of the tree. Random Forest is a bunch of decision trees created from several samples that are combined by using the majority vote for classification tasks. It is an ensemble strategy that combines the goodness of several classifier trees. For this reason, random forests are anticipated to perform better than decision trees.

3.2   Detection Using 3DCNN

Deep Learning (DL) is a subclass of AI that attempts to build neural networks and imitate the learning mechanism of the human brain. 2-dimensional CNNs are used to process normal image data. Unlike 2DCNNs, we architectured a 3DCNN such that 3-dimensional volumetric data is processed and extrapolated into their latent representations using a series of 3d convolution, 3d maximum pooling, and batch normalization operations. Three-dimensional augmentation is also applied so as to capture maximum features from each data cube.

A CNN is an intelligent architecture that specifically captures spatial features from two-dimensional image data. This is performed using the convolution operation that applies a convolution filter on the input data and produces a feature map. Each pixel in the feature map would be the weighted sum of the masked image area over which the filter is placed. The convolution operation recognizes high-dimensional features such as corners, edges, and textures in the image. Repeated convolution extracts more latent and hidden features by digging deep into images.

Fig. 3. Architecture of the proposed 3DCNN

Max pooling further decreases the size of the feature maps by selecting only the prominent feature value (by averaging or taking the maximum of values) from the pooling window. The process repeats to get the most confined representation of the image. Batch normalization is a technique for standardizing network inputs considering mini-batches of data. This helps in decreasing the run time complexity of the model. For the purpose of supervised classification, a classifier is attached as the final layer which in most cases is a fully connected layer of neurons.

We propose the architecture illustrated in Fig. 3 by converting all the aforementioned 2-dimensional operations into 3-dimensional. The layers were arranged in such a way that the cube size reduces by half after each set of Convolution, Max Pooling, and Batch Normalization operations. This is achieved by setting a 3×3 kernel-sized filter for convolution and a 2×2 pooling size filter. The classifier used here is a Multi-Layer Perceptron with a pair of dense layers. Global average pooling computes the average for each filter map. Likewise, we obtain 32 values, one for each of the feature maps from the last operation set. They are then passed through a layer comprising 512 neurons, and a second layer consisting of only 4 neurons intended for classification. The final softmax layer in the model generates a vector of predicted probability distributions and is further compared with the real distribution. The observations are evaluated using the accuracy, sensitivity, and specificity metrics.

4   Results and Discussion

The models were scripted in Python 3.8.5 and executed on an NVIDIA GeForce MX350 GPU. All of the experiments were carried out with the machine learning classifiers available in the scikit.learn package and the deep learning framework available in the Tensorflow platform v2.6.0.

We performed multiple trials with different settings of the machine learning algorithms. The number of neighbours 'K' in KNN varied from 3 to 7. The depth of the tree algorithms was set to 20 and 5 respectively through the trial and error method. Moreover, the linear Support Vector Classifier was chosen with the maximum iteration parameter set to 1000, and Logistic Regression was employed with the default settings.

The 3DCNN was optimized with the *Adam* optimizer set with the default configurations. The learning rate was initially set at 0.001 that could reduce to a minimum of 0.00001 whenever the validation loss did not improve for 3 continuous epochs. The classification error was calculated using the Categorical Cross Entropy cost function. It compares and calculates the difference between the two probability distributions (actual and forecasted). We regularized the models by artificially augmenting each data cube during training.

These models were evaluated using the Accuracy, Sensitivity, and Specificity metrics, which were calculated using a confusion matrix shown for the predicted lesions vs the actual lesion classes (Eqs. 1 through 3). The multiple blocks to which a predicted label may belong are depicted in a confusion matrix. True positives and negatives relate to how many lesion classes were successfully predicted, whereas false positives and negatives refer to how many were wrongly predicted. Accuracy computes the ratio of rightly predicted versus the entire dataset. Sensitivity evaluates the model's ability to accurately label points tagged from a specific class. Specificity, also known as the true negative rate, assesses the model's ability to identify data points that do not have the class characteristic.

$$Accuracy = \frac{1}{N}(T_P + T_N) \tag{1}$$

$$Sensitivity = \frac{T_P}{T_P + F_N} \tag{2}$$

$$Specificity = \frac{T_N}{T_N + F_P} \tag{3}$$

4.1   Experimental Results

Table 2 illustrates the accuracy of training and testing machine learning models on flattened CT data cubes. It is observed that, though most of the models fit perfectly on the training data with exceedingly high training accuracies, they fail the testing phase due to their incapability in capturing the intrinsic features of the data cube. Logistic regression and SVM are by far the best of all machine learning algorithms attaining a test accuracy of 79.52% and 77.11% respectively. Furthermore, Random forest also attempted comparable results with a test accuracy of 73.49%.

Table 2. Medical CTGAN detection using machine learning algorithms

Classifier	Settings	TrainAccuracy (%)	TestAccuracy (%)
kNN	k = 3	82.01	55.42
kNN	k = 5	70.60	57.83
Decision Tree	Max depth = 20	99.33	54.22
Random Forest	Max depth = 5	99.33	73.49
SVC linear	max iterations = 1000	**99.33**	**77.11**
Logistic regression	–	**99.33**	**79.52**

Table 3. Comparative performance (in %) of varied batch sizes in the proposed 3DCNN

Model	Accuracy	Sensitivity	Specificity
Batch size = 32			
3DCNN+Dense (256)	81.93	82.16	94.02
3DCNN+Dense (512)	85.54	85.35	95.19
Batch size = 16			
3DCNN+Dense (4)	77.11	76.88	92.33
3DCNN+Dense (128)	81.92	81.77	93.98
3DCNN+Dense (256)	83.13	82.90	94.41
3DCNN+Dense (512)	87.95	87.67	95.99
Batch size = 8			
3DCNN+Dense (128)	83.13	82.97	94.39
3DCNN+Dense (256)	86.75	86.48	95.56
3DCNN+Dense (512)	**91.57**	**91.42**	**97.20**

The second set of observations was evaluated by training and testing the 3DCNN deep neural architecture using the train, validation, and test splits of the whole dataset. The network was trained for a total of 150 epochs with a batch size varying from 8 to 32. The net number of neurons in the pre-final dense layer was also altered to analyze the effects of latent distributions in the fully connected classification layers. Table 3 projects the observed metrics of the various experiments conducted with varying batch sizes and dense neurons.

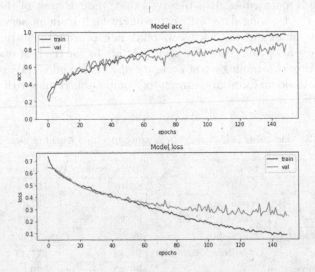

Fig. 4. Accuracy and loss curves of training and validation phases

We observed that irrespective of the batch size, better representations of the features from the CNN were interpolated using classifiers with denser layers. Decreasing the batch size also had an inverse effect on the performance of the framework. Interestingly, each of the 3DCNN configurations performs similarly to their counterparts with doubled batch size and doubled dense neurons in the pre-final layer. For instance, 3DCNN with batch size=8 and dense_neurons=128 performs akin to the 3DCNN with batch size=16 and dense_neurons=256.

Consequently, it was perceived that the ingrained representations were expressed well at a lower batch size with a denser fully connected classifier. Specifically, The model performed best with a batch size of 8 and a dense layer with 512 neurons followed by the final layer with 4 neurons each exhibiting the prediction probabilities of the different classes, thereby learning effectively from mini-batches. Overall, the best performance obtained was remarkably 91.57% accurate, 91.42% sensitive, and 97.20% specific.

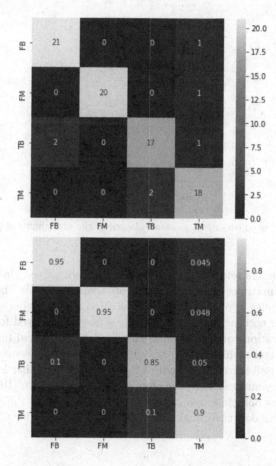

Fig. 5. Confusion matrix and normalized confusion matrix of proposed architecture

Figure 4 depicts the training and validation accuracy as well as loss curves of the best performing model. The training curve shows how well the model matches the training data, whereas the validation accuracy represents the trained model's behavior on unknown data. The validation curve was found to be consistent with the training curve, ruling out any notion that the model was over-fit to the train data. Until around 50 epochs, the curve continues to improve exponentially after which the model shows convergence by moving into the most refined state. Any more training would be futile because it would only raise the computing complexity as well as lead to over-fitting. Consequently, we chose to select the model state at the starting point of convergence (50th epoch) as the optimized final model for evaluation.

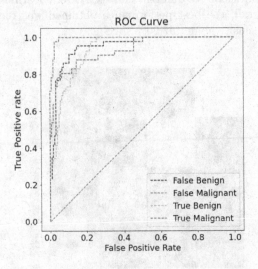

Fig. 6. ROC based on one-against-all classes of the classification predictions

The true numbers and the proportion of the categorized classes are shown in the confusion matrices (normalized and otherwise) [Fig. 5]. The fake injected and removed nodules were observed to be detected with a higher specificity than the real data. An accuracy of approximately 95% was achieved for the fake test data. Misclassifications of the fake classes were labeled as real malignant cases. In a practical setting, this may cause serious outcomes. However, mislabelling real as fake may still leave the benefit of the doubt as to further looking into the case. Figure 6 presents the Receiver Operating Characteristic (ROC) curves for the predictions as one class against all classes computed at different classification levels. The model performed with an average Area Under Curve (AUC) of 95.27%.

Table 4. Comparative performance of the proposed architecture in 2-dimensional and 3-dimensional setting

Model	Accuracy	Sensitivity	Specificity
2DCNN+Optimized Setting	87.95	87.61	95.98
Our Model	**91.57**	**91.42**	**97.20**

Further, we experimented with the same architecture in the 2-dimensional setting where convolution and max-pooling are 2D operations. The slice consisting of the central coordinate of tampering (z, x, y) is chosen to train and test the model. Specifically, the fourth slice in each datacube was selected for model tuning. Table 4 compares the model performance with the optimized 3DCNN model setting of a denser fully connected layer with 512 neurons and a batch size of 8. A clear performance gain of $\approx 4\%$ is observed. This shows that the temporal feature extraction across multiple slices performed by 3DCNN stands more significant compared to the spatial content learning of its 2D counterpart.

5 Conclusion

This study investigates the application of 3DCNN for 3D medical deepfake detection. The architecture is evaluated on the CT-GAN dataset comprising artificially synthesized malignant and benign nodules. We treat this as a multi-classification problem, as opposed to the standard deepfake detection task, which is a binary classification problem. Localized data cubes with the afflicted regions are sectioned from the CT slices and categorized using the 3DCNN. The performance of the 2-dimensional counterpart of 3DCNN is also compared. Furthermore, the performance of raw data using machine learning techniques is also assessed. This topic has garnered much too little attention. Given the potential for medical deepfakes to become a future weapon, there is huge scope for research in the domain. More complicated architectures capable of capturing inherent and latent dynamics of data might be employed to improve and generalize detection in the future.

Code and Data availability: http://mirworks.in/downloads.php

Acknowledgements. The authors would like to thank all members of the MIR (Machine Intelligence Research) group, who have immensely provided all support during the various stages of the study. The authors also thank University of Kerala for providing the infrastructure required for the study.

References

1. Goodfellow, I., et al.: Generative adversarial nets. In: Advances in Neural Information Processing Systems, vol. 27 (2014)

2. Suwajanakorn, S., Seitz, S.M., Kemelmacher-Shlizerman, I.: Synthesizing obama: learning lip sync from audio. ACM Trans. Graph. (ToG) **36**(4), 1–13 (2017)
3. Dolhansky, B., et al.: The deepfake detection challenge (dfdc) dataset. arXiv preprint arXiv:2006.07397 (2020)
4. Ciftci, U.A., Demir, I., Yin, L.: Fakecatcher: Detection of synthetic portrait videos using biological signals. IEEE Trans. Pattern Anal. Mach. Intell. (2020)
5. Siegel, D., Kraetzer, C., Seidlitz, S., Dittmann, J.: Media forensics considerations on deepfake detection with hand-crafted features. J. Imaging **7**(7), 108 (2021)
6. Mittal, T., Bhattacharya, U., Chandra, R., Bera, A., Manocha, D.: Emotions don't lie: An audio-visual deepfake detection method using affective cues. In: Proceedings of the 28th ACM International Conference on Multimedia, pp. 2823–2832 (2020)
7. Heo, Y.J., Yeo, W.H., Kim, B.G.: Deepfake detection algorithm based on improved vision transformer. Appli. Intell. 1–16 (2022)
8. Pu, J., Mangaokar, N., Wang, B., Reddy, C.K., Viswanath, B.: Noisescope: detecting deepfake images in a blind setting. In: Annual Computer Security Applications Conference, pp. 913–927 (2020)
9. Thambawita, V., et al.: Deepfake electrocardiograms using generative adversarial networks are the beginning of the end for privacy issues in medicine. Sci. Rep. **11**(1), 1–8 (2021)
10. Frid-Adar, M., Diamant, I., Klang, E., Amitai, M., Goldberger, J., Greenspan, H.: Gan-based synthetic medical image augmentation for increased cnn performance in liver lesion classification. Neurocomputing **321**, 321–331 (2018)
11. Ghorbani, A., Natarajan, V., Coz, D., Liu, Y.: Dermgan: Synthetic generation of clinical skin images with pathology. In: Machine Learning for Health Workshop, pp. 155–170. PMLR (2020)
12. Han, C., et al.: Combining noise-to-image and image-to-image gans: Brain mr image augmentation for tumor detection. IEEE Access **7**, 156966–156977 (2019)
13. Levine, A.B., et al.: Synthesis of diagnostic quality cancer pathology images by generative adversarial networks. J. Pathol. **252**(2), 178–188 (2020)
14. Cao, B., Zhang, H., Wang, N., Gao, X., Shen, D.: Auto-gan: self-supervised collaborative learning for medical image synthesis. In: Proceedings of the AAAI Conference on Artificial Intelligence, vol. 34, pp. 10486–10493 (2020)
15. Mangaokar, N., Pu, J., Bhattacharya, P., Reddy, C.K., Viswanath, B.: Jekyll: Attacking medical image diagnostics using deep generative models. In: 2020 IEEE European Symposium on Security and Privacy (EuroS&P), pp. 139–157. IEEE (2020)
16. Mirsky, Y., Mahler, T., Shelef, I., Elovici, Y.: {CT-GAN}: Malicious tampering of 3d medical imagery using deep learning. In: 28th USENIX Security Symposium (USENIX Security 2019), pp. 461–478 (2019)
17. Solaiyappan, S., Wen, Y.: Machine learning based medical image deepfake detection: A comparative study. Mach. Learn. Appli. **8**, 100298 (2022)
18. Armato, S.G., III., et al.: The lung image database consortium (lidc) and image database resource initiative (idri): a completed reference database of lung nodules on ct scans. Med. Phys. **38**(2), 915–931 (2011)

A Study on the Autonomous Detection of Impact Craters

Nour Aburaed[1]([⊠]) [iD], Mina Alsaad[1] [iD], Saeed Al Mansoori[2] [iD],
and Hussain Al-Ahmad[1] [iD]

[1] University of Dubai, Dubai, UAE
nour.aburaed@ieee.org
[2] Mohammed Bin Rashid Space Centre, Dubai, UAE

Abstract. Planet surface studies is one of the most popular research areas in planetary science, as it is useful to attain information about a planet's history and geology without directly landing on its surface. Autonomous detection of craters has been of particular interest lately, especially for Mars and Lunar surfaces. This review study deals with the technical implementation, training, and testing of YOLOv5 and YOLOv6 to gauge their efficiency in detecting craters. YOLOv6 is the most recent member of the YOLO family, and it is believed that it outperform all of its predecessors. In addition to comparing the aforementioned two models, the performance of the most widely used optimization functions, including SGD, Adam, and AdamW is studied as well. The methods are evaluated using mAP and mAR to verify whether YOLOv6 potentially outperforms YOLOv5, and whether AdamW is capable to generalize better than its peer optimizers.

Keywords: Object detection · Convolutional neural networks · YOLO · Craters · Optimizers

1 Introduction

Exploring the solar system has been a popular research interest since time immemorial. With the rapid and revolutionary improvement in technology that enables scientists to collect data and interact with outer space, the interest in this area of research continues to increase steadily throughout the years and across several industrial prospects. Mars has been the most studied planet after Earth due to their similar geological characteristics. Furthermore, there are already large archives of Martian numerical and visual datasets [9]. Mars has various surface features that are of particular importance, such as impact craters, valleys, and sand dunes. Impact craters (or simply craters) are considered as one of the most important features on Mars as well as the moon. They are a result of collision impacts, such as meteoroids, massive asteroids, and so on with martian or lunar surfaces [50]. The interest in detecting craters rose especially after NASA launched and landed Mars Rover in 2020 to explore Mars surface, which

N. El Gayar et al. (Eds.): ANNPR 2022, LNAI 13739, pp. 181–194, 2023.
https://doi.org/10.1007/978-3-031-20650-4_15

uses autonomous navigation to cover larger distance. Craters present potential obstacles for autonomous navigation vehicles, and thus, scientists utilized image processing and Artificial Intelligence (AI) technology to investigate the size, shape, and density of various craters on Mars and the moon. Additionally, studying craters helps in understanding the relative age, timescale, weathering, and geology of a planet's surface without the need to land on it [22].

The manual methods for detecting and counting craters from images rely heavily on human experts to annotate craters by visual interpretation of images, which is considered labor-intensive, time consuming, and subjective to human error. Consequently, autonomous and robust crater detection is very important to develop an automatic extraction approach for detecting and identifying crater features from images with minimal human interventions. Nowadays, AI, especially Deep Learning (DL) techniques, play a vital role in object detection generally, including the analysis of Maritain and lunar surfaces [30]. Convolutional Neural Networks (CNNs) are a subset of Artificial Neural Networks (ANNs), which are in turn a subset of Machine Learning and AI. CNNs have proven their efficiency in various computer vision tasks, including segmentation [46], classification and object detection [1]. Hence, many CNN-based object detection methods are adopted to automatically extract the impact crater.

This paper presents an experimental review study on the autonomous detection of craters on Mars and lunar surfaces by comparing the performance of You Only Look Once (YOLO)-v5 and -v6 techniques. Specifically, the optimization technique is studied as a crucial hyperparameter in each network architecture. The studied optimization functions include Stochastic Gradient Descent (SGD), Adaptive Moment Estimation (Adam), and weight decay Adam (AdamW). There exists a consensus that YOLOv6 outperforms YOLOv5, and that AdamW has better ability to generalize than Adam and SGD. This consensus will be verified through this study. The efficiency is determined according to mean Average Precision (mAP) and mean Average Recall (mAR). The rest of the paper is organized as follows: Sect. 2 presents the necessary background related to object detection, YOLO, and the optimization functions, Sect. 3 showcases the dataset used in this study, Sect. 4 explains the experimental setup, Sect. 5 demonstrates the results and draws deductions based on them, finally, Sect. 6 concludes the paper by summarizing the findings and stating the future directions of this study.

2   Background

In computer vision and remote sensing, object detection is defined as the task of locating and identifying objects in an image or video by assigning the class to which this object belongs [8]. There exists a wide range of applications, such as traffic monitoring, autonomous vehicle navigation, and environmental monitoring [27]. However, object detection is considered as a challenging task due to the large variations in object scales and orientations, background complexity, and lack of training labeled samples [35]. Object detection algorithms can be categorized into traditional approaches and Deep CNNs (DCNNs)-based approaches. Some of the most successful traditional methods for object detection include template matching, knowledge-based methods, Object-based Image Analysis (OBIA), and Machine Learning (ML). Template matching is one of the simplest and earliest approaches for object detection, however, it is not intelligent as it requires hand-crafted features. It is also computationally expensive due to its dependence on sliding windows [6]. The Knowledge-based algorithms include geometric and context knowledge [13]. OBIA technique mainly relies on analyzing images at the object level instead of pixel level. This method is divided into two main steps: the first step is segmenting the image into objects, and the second one is classifying those objects using their shape, size, spatial and spectral properties [38]. ML-based detection method extracts features manually by using various feature extraction techniques, such as Histogram of Oriented Gradients (HOG) [11], Speeded Up Robust Features (SURF) [3], Local Binary Patterns (LBP) [39], Scale-Invariant Feature Transform (SIFT) [34]. After extracting features, an independent classification technique is used to discriminate the target object. Support Vector Machine (SVM) is the most common supervised classifier that was originally proposed by the authors in [48]. SVM can be considered as a linear binary classifier that efficiently identifies and detects an object from its background with the minimum training data. The main drawbacks of the conventional approaches are their instability towards various object sizes and categories, and lighting conditions. It is also considered computationally complex. Thus, these approaches were diminished gradually and replaced by DCNNs techniques [18]. Nowadays, DCNNs methods are state-of-the-art in computer vision and image processing tasks, particularly object detection problems due to the availability of robust labeled datasets and hardware with GPU and TPU processing capabilities to perform computation tasks in edge devices. In general, DCNNs-based object detection algorithms can be divided into two groups: one-stage and two-stage object detection approaches based on the model architecture. Two-stage object detectors, utilize Region Proposal Network (RPN) to generate a series of anchor-based proposal boxes and then regress the refined bounding boxes and identify the categories of objects to produce the final output. Region proposal-based detectors include spatial pyramid pooling (SPP-net) [21], Feature Pyramid Networks (FPN) [28], Region-based Fully Convolutional Network (R-FCN) [10], Region-based Convolutional Neural Networks (R-CNN) [17], Mask R-CNN [20], Cascade R-CNN [5], Fast R-CNN [16], and Faster R-CNN [42]. On the other hand, one-stage object detectors skip candidate box extraction stage and directly performs mapping from image pixels to bounding box coordinates and frequency probabilities, which effectively treats the detection task as a regression problem. It is faster and simpler when compared to two-stage detectors. Common examples of these detectors include the YOLO family [4,40,41], Single-Shot Detector (SSD) [32], Deeply Supervised Object Detector (DSOD) [45], RetinaNet [29], CenterNet [12], CornerNet [26], Deconvolutional Single Shot Detector (DSSD) [14], and EfficientDet [47].

Due to the importance of craters for planetary science, several research studies have been conducted and addressed the automatic detection of craters from images. Some of the aforementioned traditional approaches have been indeed utilize for crater detection, such as Hough transform [15], template matching [2], Canny edge detection [43], and others. Additionally, other researchers utilizes machine learning approaches for the same purpose. Some examples include random forest [52] and SVM [44]. However, the efficiency and accuracy of these approaches is limited, and now DCNNs approaches are the most active area of research. Chaudhary et al. [7] employed YOLO object detection technique for automatic detection of impact craters from images captured by Context Camera onboard the Mars Reconnaissance Orbiter. The model was trained on an augmented dataset of 1450 images and tested on 750 images. Mean Average Precision (mAP) of 78%, precision of 81%, recall of 76% and F1 score of 78% were achieved. Researchers in [23] developed an improved UNet model for extracting the crater features from Lunar Digital Elevation Model (DEM) images. In order to improve the detection accuracy, residual network and dense connection acceleration model were introduced in the model. A post processing technique based on thresholding is also used to reduce the number of false positive in the detection results. An accuracy of 93.4% was achieved through repeating training for 5000 times. Another study [24] presented Faster R-CNN for small-scale crater detection from high resolution images collected by narrow-angle camera of the lunar reconnaissance orbiter camera (LROC NAC) for landing site of Chang'E 4, rather than the whole moon. The detection algorithm is effective and outperforms traditional machine learning algorithms, such as SVM and HOG, with an accuracy of 92.9%. Wu et al. [51] introduced a compact and automatic crater detection network PRU-Net which is mainly based on DeepMoon and ERU-Net architectures to accurately recognize craters on the Moon surface. The experiments indicate that PRU-Net provide good results with small model size about 13.6% of ERU-Net size and acceptable computational cost. In [49], the authors proposed a novel CraterIDNet model that is end-to-end Full CNN for crater detection, which accepts planetary images of any size. Five High Resolution Stereo Camera (HRSC) nadir panchromatic images taken by the Mars Express spacecraft for different geographical locations under different lighting conditions and these images were used to train the detector. Transfer learning is adopted to train the model, then anchor scale optimization and anchor density adjustment are introduced for crater detection. Grid pattern layer is added to integrate the distribution and scale information of nearby craters into a grayscale pattern image, which in turn will improve the whole detection process. Experiments showed that the proposed model obtained an accuracy that exceeds 97% even in a complex terrain scene. This technique is robust against crater position and apparent diameter errors.

Recently, YOLOv5 gained a significant advantages in autonomous object detection task in terms of speed and detection accuracy, as it includes four different modules; Focus, CSPbottleneck, Spatial pyramid pooling (SPP), and Path aggregation network (PANet) [19]. This is the main reasons behind selecting YOLOv5 as the first crater's detector in this paper. First, YOLOv5 uses cross stage partial network (CSPNet) into Darknet, creating CSPDarknet as its backbone. This backbone will solve the repetitive gradient information in large scale backbones and integrates gradient change into feature map which improve the inference speed and accuracy and decreases the parameters, which in turn reduces the model size. Second, PANet is used as neck to boost the information flow. PANet adopts a new FPN that includes several bottom ups and top down layers to enhance the propagation of low level features in the model. PANet also improves the localization in lower layers, which obviously improves the localization accuracy of the object. Third, The head of YOLOv5 is exactly the same as YOLOv4 and YOLOv3 heads that generates three different sizes (18×18, 36×36, 72×72) of feature maps to achieve multi-scale prediction that enable the model to deal with small, medium, and big objects [37]. The second autonomous detector chosen for this study is the most recent version of the YOLO family, which is YOLOv6, a single-stage object detection framework dedicated to industrial applications. This model is not part of the official YOLO series, but according to [36], it outperforms YOLOv5 in terms detection accuracy and inference speed, which is more efficient for industrial applications. This study tests this claim by deploying both YOLOv5 and YOLOv6 to detect craters using SGD, Adam, and AdamW as optimizers.

3   Dataset

The dataset used in this paper is the Martian/Lunar Crater Detection Dataset provided by Kaggle [31]. The dataset mainly contains images of Mars and Moon surface which may contain crater features. The data is from different sources. For Mars, the images are mainly from Arizona State University (ASU) and United States Geological Survey (USGS), while all Moon images are from NASA Lunar Reconnaissance Orbiter mission. All images are pre-processed with RoboFlow to remove EXIF rotation, and they are resized to 640×640. The annotation work was performed by a group from Tongji University, Shanghai, China in YOLO text format. Samples from the dataset are shown in Fig. 1.

4   Experimental Setup

4.1   Optimization Functions

Optimization algorithms are used to minimize the loss of predictive models, such as DCNNs, with respect to a certain training dataset by adjusting the model's parameters, such as the weight and/or learning rate. Detection of craters can be can considered as an object detection problem that consists of only one class of objects. For an ith sample of a crater object denoted x with label y, SGD optimizer is defined as:

$$\Theta = \Theta - \alpha \cdot \nabla_\Theta J(\Theta; x^{(i)}; y^{(i)}) \tag{1}$$

such that Θ represents the model's parameters, $\nabla_\Theta J$ is the objective function, and α is the learning rate. SGD's advantage is that it is generally fast, and its

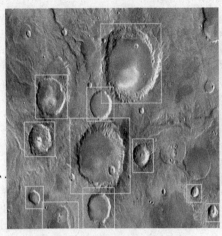

(a) Ground truth for lunar image   (b) Ground truth for mars image

Fig. 1. Sample from mars/lunar crater dataset

fluctuations enable it to potentially escape from one local minimum to a better one, but it makes finding the global minimum more complicated.

Adam is another optimization algorithm that was first proposed in [25]. It divides the learning rate by an exponentially decaying average of squared gradients, which is an idea that was first seen in earlier optimization functions, such as RMSProp and Adadelta. Additionally, Adam uses a mechanism known as *momentum*, such that it keeps an exponentially decaying average of past gradients. This gives Adam greater advantage in terms of adaptive learning. The decaying averages at a certain time step t are computed as follows:

$$\hat{m}_t = \frac{\beta_1 m_{t-1} + (1 - \beta_1) g_t}{i - \beta_1^t} \tag{2}$$

$$\hat{v}_t = \frac{\beta_2 v_{t-1} + (1 - \beta_2) g_t^2}{1 - \beta_2^t} \tag{3}$$

g_t is the gradient at time t, and m_t and v_t are the estimates of mean and uncentered variance, respectively. Consequently, Adam updates the DCNN's parameters as follows:

$$\Theta_{t+1} = \Theta_t - \frac{\alpha}{\sqrt{\hat{v}} + \epsilon} \hat{m}_t \tag{4}$$

The most recent version of Adam is AdamW [33]. Loshchilov and Hutter argue that Adam does not have the ability to generalize, and they solve this issue by controlling the weight decay, such that it is performed after controlling the parameter-wise step size. AdamW is expressed as follows:

Table 1. Training parameters for YOLOv5 and YOLOv6 experiments.

Parameter	Value
Epochs	500
Loss	MSE
Optimization	SGD/Adam/AdamW
Scheduler	Cosine annealing

$$\Theta_{t+1} = \Theta_t - \alpha \frac{\beta_1 - m_t + (1 - \beta_1)(g_t + w\Theta_t)}{\sqrt{\hat{v}} + \epsilon} \tag{5}$$

4.2 Training Strategy

The dataset discussed in Sect. 3 is divided into 98 images for training, 26 for validation, and 19 for testing. In order to ensure fair comparison, the same training parameters are applied to both YOLOv5 and YOLOv6 across all experiments. Furthermore, an early stopping strategy is deployed, such that if the network does not improve for 100 consecutive epochs, then it stops training prematurely. Also, the evaluation is performed based on the best parameters obtained, and not based on the most recent epoch. The training parameters are summarized in Table 1.

5 Results

Whether a detection is considered as True Positive (TP), True Negative (TN), False Positive (FP), or False Negative (FN), this is decided depending on the Intersection Over Union (IoU), which assesses the extent of overlap between the detected bounding box and the groundtruth bounding box. IoU is defined as follows:

$$IoU = \frac{Area\,of\,overlap}{Area\,of\,union} \tag{6}$$

For this study, the threshold value of IoU is 0.5, such that if the detected bounding box overlaps with the groundtruth one by at least 50%, then the detection is considered as TP. Detections with less than 50%, where 0% indicates failure to detect the object altogether, are depicted as FN. If the model predicts a crater where no crater exists, the case is regarded as FP, and if it successfully avoids this prediction, then it is a TP case.

The results of the experiments are evaluated using mAP and mAR. Precision indicates the model's ability to find TP among all TP and FP cases. mAP measures precision across all classes c. Since there is only one class in this study, mAP and precision are calculated the same way and can be used interchangeably, such that:

$$mAP = P = \frac{1}{c} \sum_{i=1}^{i=c} \frac{TP}{TP+FP} \tag{7}$$

Recall indicates the model's ability to find TP among all TP and FN cases. Similar to mAP, mAR measures recall across all classes, such that

$$mAR = R = \frac{1}{c} \sum_{i=1}^{i=c} \frac{TP}{TP+FN} \tag{8}$$

Table 2 summarizes the results of the experiments in terms of mAP and mAR. Overall, YOLOv5 shows better performance than YOLOv6 in terms of mAP. At the same time, YOLOv6 shows better performance than YOLOv5 in terms of mAR. AdamW shows better performance compared to other optimization algorithms in both YOLOv5 and YOLOv6 for mAP, but not for mAR. AdamW shows the best mAP with YOLOv5, while SGD shows the best mAR for the same network. For YOLOv6, Adam shows the best mAP and mAR. It is worth mentioning that there are cases where YOLOv6 with AdamW successfully detect craters in complicated scenes, whereas it fails to do so with other optimization functions, and YOLOv5 generally fails to detect them as well. Figure 2 depicts such case, where Figs. 2(a-e) all show FN, FP, and redundant detections, which are also regarded as FP. On the other hand, Fig. 2(f), does not show such cases. Conversely, there are cases where YOLOv6 with SGD shows the best result, as seen in Figre 3.

It seems that despite the strengths of each model and optimization function, there is still no optimal way to avoid all the shortcomings and cover all the scenarios, as each model and each optimization function are successful in specific cases. This opens room for improvement and draws a new direction in this area of research. Furthermore, YOLOv6 did not outperform YOLOv5 in all scenarios, and AdamW did not outperform Adam and SGD in all scenarios, which contradicts the general belief about YOLOv6 and AdamW.

Table 2. Results summary

Model	SGD		Adam		AdamW	
	mAP	mAR	mAP	mAR	mAP	mAR
YOLOv5	**0.72**	0.54	**0.73**	0.48	**0.74**	0.49
YOLOv6	0.62	**0.64**	0.62	**0.65**	0.55	**0.47**

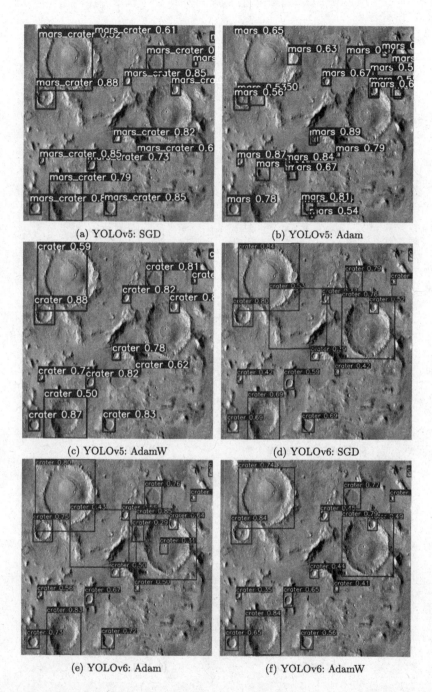

(a) YOLOv5: SGD

(b) YOLOv5: Adam

(c) YOLOv5: AdamW

(d) YOLOv6: SGD

(e) YOLOv6: Adam

(f) YOLOv6: AdamW

Fig. 2. Sample results from YOLOv5 with (a) SGD, (b) Adam, and (c) Adam w, and sample results from YOLOv6 with (d) SGS, (e) Adam, (f) AdamW. YOLOv6 with AdamW shows the best results.

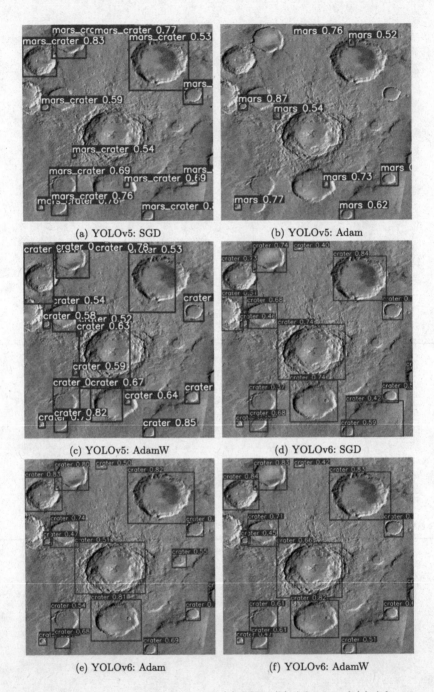

(a) YOLOv5: SGD   (b) YOLOv5: Adam

(c) YOLOv5: AdamW   (d) YOLOv6: SGD

(e) YOLOv6: Adam   (f) YOLOv6: AdamW

Fig. 3. Sample results from YOLOv5 with (a) SGD, (b) Adam, and (c) Adam w, and sample results from YOLOv6 with (d) SGS, (e) Adam, (f) AdamW. YOLOv6 with AdamW shows the best results.

6 Conclusion and Future Work

In this study, an experimental review of YOLOv5 and YOLOv6 models with SGD, Adam, and AdamW optimization functions has been conducted using have been trained over Martian/Lunar Crater Detection Dataset. The general belief is that YOLOv6 is an upgraded version of YOLOv5, and that AdamW optimizer outperforms SGD and Adam. The networks and optimization functions are tested systematically to verify these claims. Experiments show inconsistent outcomes, as there are cases where YOLOv5 performs better than YOLOv6, and vice versa, and cases where each optimization function outshines others under different conditions. In the future, YOLOv6 model will be extended to the wavelet domain in order to detect objects using frequency analysis in order to avoid the inconsistencies and shortcomings faced in its current performance. Also, the model will be trained and tested using Mars data collected by Emirates Hope Probe.

References

1. Aburaed, N., Al-Saad, M., Chendeb El Rai, M., Al Mansoori, S., Al-Ahmad, H., Marshall, S.: Autonomous object detection in satellite images using WFRCNN. In: 2020 IEEE India Geoscience and Remote Sensing Symposium (InGARSS), pp. 106–109 (2020). https://doi.org/10.1109/InGARSS48198.2020.9358948
2. Bandeira, L.ç., Saraiva, J., Pina, P.: Impact crater recognition on mars based on a probability volume created by template matching. IEEE Trans. Geosci. Remote Sens. **45**(12), 4008–4015 (2007). https://doi.org/10.1109/TGRS.2007.904948
3. Bay, H., Ess, A., Tuytelaars, T., Van Gool, L.: Speeded-up robust features (SURF). Comput. Vis. Image Underst. **110**(3), 346–359 (2008)
4. Bochkovskiy, A., Wang, C.Y., Liao, H.Y.M.: Yolov4: optimal speed and accuracy of object detection. arXiv preprint arXiv:2004.10934 (2020)
5. Cai, Z., Vasconcelos, N.: Cascade R-CNN: delving into high quality object detection. In: Proceedings of the IEEE Conference on Computer Vision and Pattern Recognition, pp. 6154–6162 (2018)
6. Chantara, W., Ho, Y.S.: Object detection based on fast template matching through adaptive partition search. In: 2015 12th International Joint Conference on Computer Science and Software Engineering (JCSSE), pp. 1–6. IEEE (2015)
7. Chaudhary, V., Mane, D., Anilkumar, R., Chouhan, A., Chutia, D., Raju, P.: An object detection approach to automatic crater detection from CTX imagery. Technical report, Copernicus Meetings (2020)
8. Cheng, G., Han, J.: A survey on object detection in optical remote sensing images. ISPRS J. Photogramm. Remote. Sens. **117**, 11–28 (2016)
9. Christoff, N., Manolova, A., Jorda, L., Mari, J.L.: Morphological crater classification via convolutional neural network with application on MOLA data. In: ANNA 2018, Advances in Neural Networks and Applications 2018, pp. 1–5 (2018)
10. Dai, J., Li, Y., He, K., Sun, J.: R-FCN: object detection via region-based fully convolutional networks. Adv. Neural Inf. Process. Syst. **29**, 1–13 (2016)
11. Dalal, N., Triggs, B.: Histograms of oriented gradients for human detection. In: 2005 IEEE Computer Society Conference on Computer Vision and Pattern Recognition (CVPR 2005), vol. 1, pp. 886–893. IEEE (2005)

12. Duan, K., Bai, S., Xie, L., Qi, H., Huang, Q., Tian, Q.: Centernet: keypoint triplets for object detection. In: Proceedings of the IEEE/CVF International Conference on Computer Vision, pp. 6569–6578 (2019)
13. Fatima, S.A., Kumar, A., Pratap, A., Raoof, S.S.: Object recognition and detection in remote sensing images: a comparative study. In: 2020 International Conference on Artificial Intelligence and Signal Processing (AISP), pp. 1–5. IEEE (2020)
14. Fu, C.Y., Liu, W., Ranga, A., Tyagi, A., Berg, A.C.: DSSD: deconvolutional single shot detector. arXiv preprint arXiv:1701.06659 (2017)
15. Galloway, M.J., Benedix, G.K., Bland, P.A., Paxman, J., Towner, M.C., Tan, T.: Automated crater detection and counting using the Hough transform. In: 2014 IEEE International Conference on Image Processing (ICIP), pp. 1579–1583 (2014). https://doi.org/10.1109/ICIP.2014.7025316
16. Girshick, R.: Fast R-CNN. In: Proceedings of the IEEE International Conference On Computer Vision, pp. 1440–1448 (2015)
17. Girshick, R., Donahue, J., Darrell, T., Malik, J.: Rich feature hierarchies for accurate object detection and semantic segmentation. In: Proceedings of the IEEE Conference on Computer Vision and Pattern Recognition, pp. 580–587 (2014)
18. Guo, J.M., Yang, J.S., Seshathiri, S., Wu, H.W.: A light-weight CNN for object detection with sparse model and knowledge distillation. Electronics 11(4), 575 (2022)
19. Guo, Z., Wang, C., Yang, G., Huang, Z., Li, G.: MSFT-YOLO: improved YOLOV5 based on transformer for detecting defects of steel surface. Sensors 22(9), 3467 (2022)
20. He, K., Gkioxari, G., Dollár, P., Girshick, R.: Mask R-CNN. In: Proceedings of the IEEE International Conference on Computer Vision, pp. 2961–2969 (2017)
21. He, K., Zhang, X., Ren, S., Sun, J.: Spatial pyramid pooling in deep convolutional networks for visual recognition. IEEE Trans. Pattern Anal. Mach. Intell. 37(9), 1904–1916 (2015)
22. Hsu, C.Y., Li, W., Wang, S.: Knowledge-driven GEOAI: Integrating spatial knowledge into multi-scale deep learning for mars crater detection. Remote Sens. 13(11) (2021). https://doi.org/10.3390/rs13112116, https://www.mdpi.com/2072-4292/13/11/2116
23. Jia, Y., Wan, G., Liu, L., Wu, Y., Zhang, C.: Automated detection of lunar craters using deep learning. In: 2020 IEEE 9th Joint International Information Technology and Artificial Intelligence Conference (ITAIC), vol. 9, pp. 1419–1423. IEEE (2020)
24. Jin, Y., He, F., Liu, S., Tong, X.: Small scale crater detection based on deep learning with multi-temporal samples of high-resolution images. In: 2019 10th International Workshop on the Analysis of Multitemporal Remote Sensing Images (MultiTemp), pp. 1–4. IEEE (2019)
25. Kingma, D.P., Ba, J.: Adam: a method for stochastic optimization. In: Bengio, Y., LeCun, Y. (eds.) 3rd International Conference on Learning Representations, ICLR 2015, San Diego, CA, USA, 7–9 May 2015, Conference Track Proceedings (2015). http://arxiv.org/abs/1412.6980
26. Law, H., Deng, J.: CornerNet: detecting objects as paired keypoints. Int. J. Comput. Vision 128(3), 642–656 (2019). https://doi.org/10.1007/s11263-019-01204-1
27. Li, K., Cao, L.: A review of object detection techniques. In: 2020 5th International Conference on Electromechanical Control Technology and Transportation (ICECTT), pp. 385–390. IEEE (2020)
28. Lin, T.Y., Dollár, P., Girshick, R., He, K., Hariharan, B., Belongie, S.: Feature pyramid networks for object detection. In: Proceedings of the IEEE Conference on Computer Vision and Pattern Recognition, pp. 2117–2125 (2017)

29. Lin, T.Y., Goyal, P., Girshick, R., He, K., Dollár, P.: Focal loss for dense object detection. In: Proceedings of the IEEE International Conference on Computer Vision, pp. 2980–2988 (2017)

30. Lin, X., et al.: Lunar crater detection on digital elevation model: a complete workflow using deep learning and its application. Remote Sens. **14**(3), 621 (2022). https://doi.org/10.3390/rs14030621, https://www.mdpi.com/2072-4292/14/3/621

31. LincolnZh: Martian/lunar crater detection dataset, February 2022. https://www.kaggle.com/datasets/lincolnzh/martianlunar-crater-detection-dataset

32. Liu, W., et al.: SSD: single shot multibox detector. In: Leibe, B., Matas, J., Sebe, N., Welling, M. (eds.) ECCV 2016. LNCS, vol. 9905, pp. 21–37. Springer, Cham (2016). https://doi.org/10.1007/978-3-319-46448-0_2

33. Loshchilov, I., Hutter, F.: Decoupled weight decay regularization. In: International Conference on Learning Representations (2019). https://openreview.net/forum?id=Bkg6RiCqY7

34. Lowe, D.G.: Object recognition from local scale-invariant features. In: Proceedings of the Seventh IEEE International Conference on Computer Vision, vol. 2, pp. 1150–1157. IEEE (1999)

35. Mahmoud, A., Mohamed, S., El-Khoribi, R., Abdelsalam, H.: Object detection using adaptive mask R-CNN in optical remote sensing images. Int. J. Intell. Eng. Syst **13**(1), 65–76 (2020)

36. meituan: Yolov6, June 2022. https://github.com/meituan/YOLOv6

37. Nepal, U., Eslamiat, H.: Comparing Yolov3, Yolov4 and Yolov5 for autonomous landing spot detection in faulty UAVs. Sensors **22**(2), 464 (2022)

38. Norman, M., Shahar, H.M., Mohamad, Z., Rahim, A., Mohd, F.A., Shafri, H.Z.M.: Urban building detection using object-based image analysis (OBIA) and machine learning (ML) algorithms. In: IOP Conference Series: Earth and Environmental Science, vol. 620, p. 012010. IOP Publishing (2021)

39. Ojala, T., Pietikainen, M., Maenpaa, T.: Multiresolution gray-scale and rotation invariant texture classification with local binary patterns. IEEE Trans. Pattern Anal. Mach. Intell. **24**(7), 971–987 (2002)

40. Redmon, J., Divvala, S., Girshick, R., Farhadi, A.: You only look once: unified, real-time object detection. In: Proceedings of the IEEE Conference on Computer Vision and Pattern Recognition, pp. 779–788 (2016)

41. Redmon, J., Farhadi, A.: Yolov3: an incremental improvement. arXiv preprint arXiv:1804.02767 (2018)

42. Ren, S., He, K., Girshick, R., Sun, J.: Faster R-CNN: towards real-time object detection with region proposal networks. Adv. Neural Inf. Process. Syst. **28**, 1–10 (2015)

43. Saheba, S.M., Upadhyaya, T.K., Sharma, R.K.: Lunar surface crater topology generation using adaptive edge detection algorithm. IET Image Proc. **10**(9), 657–661 (2016). https://doi.org/10.1049/iet-ipr.2015.0232

44. Salamunićcar, G., Lončarić, S.: Application of machine learning using support vector machines for crater detection from Martian digital topography data. In: 38th COSPAR Scientific Assembly, vol. 38, p. 3, January 2010

45. Shen, Z., Liu, Z., Li, J., Jiang, Y.G., Chen, Y., Xue, X.: DSOD: learning deeply supervised object detectors from scratch. In: Proceedings of the IEEE International Conference on Computer Vision, pp. 1919–1927 (2017)

46. Talal, M., Panthakkan, A., Mukhtar, H., Mansoor, W., Almansoori, S., Ahmad, H.A.: Detection of water-bodies using semantic segmentation. In: 2018 International Conference on Signal Processing and Information Security (ICSPIS), pp. 1–4 (2018). https://doi.org/10.1109/CSPIS.2018.8642743

47. Tan, M., Pang, R., Le, Q.V.: Efficientdet: scalable and efficient object detection. In: Proceedings of the IEEE/CVF Conference on Computer Vision and Pattern Recognition, pp. 10781–10790 (2020)

48. Vapnik, V.: Statistical Learning Theory New York. Wiley, New York (1998)

49. Wang, H., Jiang, J., Zhang, G.: CrateridNet: an end-to-end fully convolutional neural network for crater detection and identification in remotely sensed planetary images. Remote Sens. **10**(7), 1067 (2018)

50. Wang, J., et al.: Effective classification for crater detection: a case study on mars. In: 9th IEEE International Conference on Cognitive Informatics (ICCI 2010), pp. 688–695 (2010). https://doi.org/10.1109/COGINF.2010.5599824

51. Wu, Y., Wan, G., Liu, L., Wei, Z., Wang, S.: Intelligent crater detection on planetary surface using convolutional neural network. In: 2021 IEEE 5th Advanced Information Technology, Electronic and Automation Control Conference (IAEAC), vol. 5, pp. 1229–1234. IEEE (2021)

52. Yin, J., Li, H., Jia, X.: Crater detection based on gist features. IEEE J. Sel. Top. Appl. Earth Observ. Remote Sens. **8**(1), 23–29 (2015). https://doi.org/10.1109/JSTARS.2014.2375066

Utilization of Vision Transformer for Classification and Ranking of Video Distortions

Nouar AlDahoul[1]([✉]) [iD], Hezerul Abdul Karim[1] [iD], and Myles Joshua Toledo Tan[2] [iD]

[1] Faculty of Engineering, Multimedia University, Cyberjaya, Malaysia
nouar.aldahoul@live.iium.edu.my, hezerul@mmu.edu.my
[2] Departments of Natural Sciences, Chemical Engineering, and Electronics Engineering,
University of St., La Salle, Bacolod, Philippines
mj.tan@usls.edu.ph

Abstract. The perceptual quality of video surveillance footage has impacts on several tasks involved in the surveillance process, such as the detection of anomalous objects. The videos captured by a camera are prone to various distortions such as noise, smoke, haze, low or uneven illumination, blur, rain, and compression, which affect visual quality. Automatic identification of distortions is important when enhancing video quality. Video quality assessment involves two stages: (1) classification of distortions affecting the video frames and (2) ranking of these distortions. A novel video dataset was utilized for training, validating, and testing. Working with this dataset was challenging because it included nine categories of distortions and four levels of severity. The greatest challenge was the availability of multiple types of distortions in the same video. The work presented in this paper addresses the problem of multi-label distortion classification and ranking. A vision transformer was used for feature learning. The experiment showed that the proposed solution performed well in terms of F1 score of single distortion (77.9%) and F1 score of single and multiple distortions (69.9%). Moreover, the average accuracy of level classification was 62% with an average F1 score of 61%.

Keywords: Distortion classification and ranking · Multi-label classification · Video quality assessment · Vision transformer

1 Introduction

As a whole, the objective of video distortion detection is to identify the occurrence of deformed objects in acquired video for the ultimate goal of evaluating its quality. While much attention has been given to this endeavor in recent years, owing to the ubiquity of video cameras nowadays, it is certainly not a new endeavor. The task of assessing video quality, which involves describing distorted objects and features in local regions, frames, and video sequences has already been described in the literature 20 years ago, or arguably even earlier [1]. Some 10 years prior to this, efforts to describe distortions in images have likewise been explored [2].

N. El Gayar et al. (Eds.): ANNPR 2022, LNAI 13739, pp. 195–204, 2023.
https://doi.org/10.1007/978-3-031-20650-4_16

The perceptual quality of video surveillance footage has wide-reaching impacts on several tasks involved in the surveillance process, such as the detection and tracking of anomalous objects and events. However, the environment from which the surveillance footage is acquired, the encoding process, and the storage of data can introduce distortions to the video despite the use of state-of-the-art video surveillance hardware. For this reason, the use of appropriate video post-processing [3–20], which nowadays often involves machine learning approaches [11–20], may become necessary. Moreover, of the learning approaches enumerated, several may be described as methods that involve the generation of frames to rectify distorted videos.[14–20].

However, prior to performing the requisite post-processing techniques in order to enhance the quality of the degraded surveillance footage, appropriate and robust methods for the detection and subsequent classification of distortions in the footage must first be developed. Unfortunately, while several methods to detect and classify distortions in laparoscopic video have been developed in response to the 2020 challenge of the International Conference on Image Processing (ICIP) [21–24], not many methods geared toward video distortion detection and classification specifically for surveillance applications have recently been described in the literature [25–28].

In the context of distortion detection and classification in laparoscopic videos, [21] proposed an algorithm that performs a multi-label classification and ranking of distortions in laparoscopic videos by employing a residual neural network (ResNet). [22], on the other hand, proposed distortion-specific methods that are no-reference, opinion-unaware, accurate, and at the same time, computationally inexpensive in order to classify these laparoscopic video distortions. In contrast, [23] proposed a transfer learning approach that employed a pre-trained ResNet50 architecture in order to extract features mapped by support vector machine classifiers. Finally, [24] initially used a pre-trained ResNet50 architecture to extract spatial features from the frames of the laparoscopic video and made use of the temporal features of a long short-term memory model to enhance classification.

In the context of video surveillance, [26] proposed a smart security monitoring system that comprised a module for the evaluation of data quality and the identification of distortions, and a module for the enhancement of data quality. In this method, data are pre-processed in order to enhance their quality if acquired under certain conditions. [28], on the other hand, described a much more traditional approach to evaluating the performance of real-time surveillance videos developed by Bosch Corporate Research in the early 2000s. The approach was based on [29], which was described in the late 1990s. Moreover, [25] demonstrated that some state-of-the-art metrics used in the evaluation of video quality are not suitable for video surveillance applications, thus stirring research on the quality assessment of surveillance video toward a different direction.

Another key issue in this challenge that needs to be addressed is the possibility that multiple video distortions can occur at the same time within a frame [25, 30]. Hence, it is necessary that a realistic dataset that contains a combination of surveillance videos degraded by single and multiple distortions be used when training the appropriate detection and classification algorithm. For this reason, the organizers of the 2022 ICIP challenge entitled "Video distortion detection and classification in the context of video surveillance" have come up with a dataset, which they called the Video Surveillance

Quality Assessment Dataset (VSQuAD), containing short-duration surveillance videos with several distortion types of varying levels of degradation [30].

The key contributions of this paper are as follows:

1) An accurate distortion classification model that was able to identify multiple distortions affecting the same video.
2) An accurate distortion ranking model that was able to estimate the levels of severity of distortions affecting the video for video quality assessment (VQA).
3) A single vision transformer model that was used for both distortion classification and ranking.

This paper is structured as follows: Sect. 2 describes the dataset and demonstrates the proposed solution. In Sect. 3, the experimental setup, and results are discussed. Section 4 summarizes the significance of this work and opens doors for further improvement.

2   Materials and Methods

This section describes the dataset and discusses the methods utilized in this work.

2.1   Dataset Overview

A novel surveillance video dataset for video distortion detection and classification in the context of video surveillance, called "Video Surveillance Quality Assessment Dataset" (VSQuAD) was used [30]. This dataset includes 36 1920 × 1080 RGB reference videos at a frame rate of 30 fps acquired in daylight and nightlight captured by fixed and moving cameras in indoor and outdoor scenes. Several samples are shown in Fig. 1. The videos distorted by nine single distortions and nine multiple distortions at four severity levels were 10 s long [30]. These distortions were artificially generated by algorithms. The description of distortion categories is shown in Table 1. There are 960 videos with single distortion and 640 videos with multiple distortions.

2.2   Vision Transformer – The Proposed Method

A recent deep learning (DL) model known as the vision transformer (ViT) [31] was used in this work for distortion classification and ranking. The ILSVRC-2012 ImageNet-2012 dataset, and ImageNet-21k dataset were utilized for ViT training. After training the ViT, it was fine-tuned using a small-scale VSQuAD dataset. A video frame consists of a sequence of patches that were encoded as words and subsequently inputted into a transformer encoder as illustrated in Fig. 2.

The ViT includes an encoder architecture with L blocks, alternating between multi-head self-attention (MSA) blocks and multilayer perceptron (MLP) blocks. Layer normalization (LN) is applied before each block, while residual connections are applied after each one.

The base type of the ViT model was used with 12 layers, an MLP size of 3072, and 12 heads. Here, there were a total of 86 million parameters.

Fig. 1. Samples with various distortions and intensities from the VSQuAD dataset [30].

Table 1. Description of distortion categories.

Single distortion	Multiple distortions
Noise	Noise + Defocus blur
Defocus blur	Noise + Low illumination
Motion blur	Noise + Uneven illumination
Smoke	Low illumination + Rain
Uneven illumination	Defocus blur + Compression
Low illumination	Low illumination + Noise + Rain
Rain	Noise + Uneven illumination + Compression
Haze	Rain + Uneven illumination + Motion blur
Compression	Haze + Noise + Compression

In this paper, the vision transformer was used as a pre-trained model. The objective was to extract 768 features from RGB video frames which were resized to 384×384. Two-layer MLPs, with 512 nodes each, were added as replacement for the top layers as follows:

1) Two-layer MLPs to map features to nine distortion types. The last layer used for distortion classification had nine nodes with nine sigmoid activation functions to

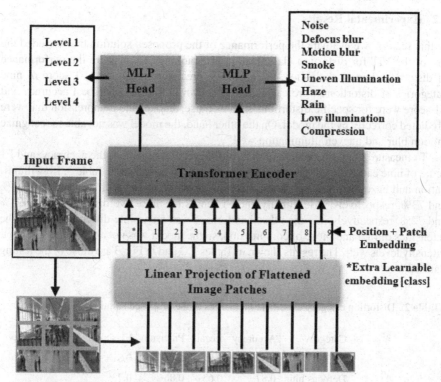

Fig. 2. The block diagram of the proposed vision transformer architecture.

produce an output of multi-label classes, such as 110010000. The loss function was a sum of nine binary cross entropy loss functions.

2) Nine two-layer MLPs to map features to four levels of intensity for each distortion were used. The last layer after each two-layer MLP used for distortion ranking had four nodes with four SoftMax activation functions for each distortion. A categorical cross entropy loss function with one-hot encoding, such as 0001, was used.

3   Results and Discussion

3.1   Experimental Setup

The experiments for training the proposed ViT architecture were conducted on an NVIDIA Tesla V100 GPU using the TensorFlow framework. The images were resized to 384×384 and applied to the input of model. ViT base 16 was selected. The hyperparameters were as follow: 10 epochs; batch size of 16; Reduce_On_Plateau learning rate with factor $= 0.1$, min_LR $= 0$, and patience $= 2$; and an Adam optimizer.

The training set consisted of 1116 RGB labeled videos. Each video was read and stored as 61 frames. Therefore, we divided the training set into three sets: 60% for training, 20% for validation, and 20% for testing.

3.2 Experimental Results

In this section, we evaluate the performance of the proposed solution that included the use of the ViT for distortion classification and ranking. To measure the performance of distortion type classification, the accuracy, precision, recall, and F1 score of nine categories of distortions were calculated as shown in Table 2. The best accuracy and F1 score were for specific distortions such as noise, smoke, and compression that were predicted correctly by the model. On the other hand, the model was not able to recognize motion blur and uneven illumination well.

To measure the performance of distortion level classification, the accuracy, and F1 score of nine categories of distortions were calculated as shown in Table 3. Low illumination and haze distortions were estimated well by the model with accuracies of 94% and 73%, respectively. Additionally, the F1 scores for these two distortions were 94% and 72%, respectively. On the other hand, there was no visible difference among the intensities of rain and those of compression, so the model was not able to estimate their intensity levels well. The results shown in Tables 2, 3 and in Fig. 3 are those of the testing dataset.

Table 2. Distortion category classification metrics of the proposed solution for each category.

Category	Accuracy	Recall	Precision	F1-score
Noise	0.84	0.94	0.7	0.81
Defocus blur	0.87	0.63	0.86	0.73
Motion blur	0.80	0.88	0.34	0.49
Smoke	0.986	0.95	0.90	0.93
Uneven illum	0.75	0.63	0.27	0.38
Compression	0.89	0.84	0.76	0.80
Haze	0.96	0.55	1.00	0.71
Low illum	0.96	0.59	1.00	0.74
Rain	0.94	0.73	0.67	0.70
Average	**0.89**	**0.75**	**0.72**	**0.70**

Figures 3 and 4 show the confusion matrix of the proposed solution for four types of distortions. The high values of the elements in the main diagonal are clear. Four confusion matrices were used to illustrate that there was good estimation of distortion intensity levels despite some videos containing multiple distortions of varying intensities. For noise distortion, 22 hardly visible, 7 visible but not annoying, 19 annoying, and 18 very annoying video distortions were classified correctly. For defocus blur, 9 hardly visible, 16 visible but not annoying, 8 annoying, and 7 very annoying video distortions were classified correctly. For uneven illumination, 8 hardly visible, 7 visible but not annoying, 11 annoying, and 11 very annoying video distortions were classified correctly. For low

Table 3. Distortion level classification metrics of the proposed solution for each category

Category	Accuracy	F1-score
Noise	0.59	0.58
Defocus blur	0.67	0.67
Motion blur	0.67	0.65
Smoke	0.55	0.46
Uneven illumination	0.66	0.66
Compression	0.41	0.40
Haze	0.73	0.72
Low illumination	0.94	0.94
Rain	0.39	0.38
Average	**0.62**	**0.61**

illumination, 9 hardly visible, 9 visible but not annoying, 7 annoying, and 9 very annoying video distortions were classified correctly.

In summary, the proposed ViT solution was a good candidate model to be used for distortion classification and ranking for surveillance purposes. More hyperparameter tuning may be necessary to boost the performance and increase the category accuracy, level accuracy, and F1 score of single and multiple distortions.

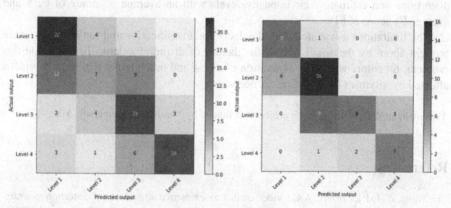

Fig. 3. Confusion matrix of proposed solution for four distortion types: noise (left), defocus blur (right).

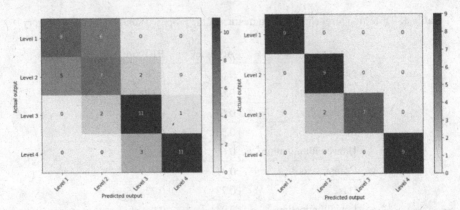

Fig. 4. Confusion matrix of proposed solution for four distortion types: uneven illumination (left), and low illumination (right).

4   Conclusion and Future Work

This paper demonstrated VQA tasks, including distortion classification and ranking for surveillance purposes. A novel dataset called VSQuAD proposed for an ICIP 2022 challenge was used. A ViT, which is a DL model utilized for learning features from video frames, was used. The experimental results showed that the proposed ViT architecture performed well in terms of F1 score of single distortion (77.9%) and F1 score of single and multiple distortion (69.9%). Similarly, the proposed solution was found to rank distortions and estimate their intensity levels with an average accuracy of 62% and average F1 score of 61%.

The limitations associated with the distortion classification and ranking tasks were brought about by the small size of the dataset of distorted videos. To overcome this problem, for future work, the generation of a new and much larger dataset with videos affected by mixtures of distortions is necessary.

Acknowledgment. This research project was funded by Multimedia University, Malaysia.

References

1. Wang, Z., Lu, L., Bovik, A.C.: Video quality assessment using structural distortion measurement. In: Proceedings. International Conference on Image Processing 2002, p. III (2002). https://doi.org/10.1109/ICIP.2002.1038904
2. Teo, P.C., Heeger, D.J.: Perceptual image distortion. In: Proceedings of 1st International Conference on Image Processing, vol. 2, pp. 982–986 (1994). https://doi.org/10.1109/ICIP.1994.413502
3. Zhang, Z.: Flexible camera calibration by viewing a plane from unknown orientations. In: Proceedings of IEEE International Conference on Computer Vision, pp. 666–673, September 1999
4. Chen, X., et al.: Calibration of a hybrid camera network. In: Proceedings of IEEE International Conference on Computer Vision, pp. 150–155, October 2003

5. Barreto, J.P., Araujo, H.: Geometric properties of central catadioptric line images and their application in calibration. IEEE Trans. Pattern Anal. Mach. Intell. **27**(8), 1327–1333 (2005)
6. Melo, R., Antunes, M., Barreto, J.P., Falcão, G., Gonçalves, N.: Unsupervised intrinsic calibration from a single frame using a 'plumb-line' approach. In: Proceedings of IEEE International Conference on Computer Vision, pp. 537–544, June 2013
7. Carroll, R., Agrawal, M., Agarwala, A.: Optimizing content-preserving projections for wide-angle images. ACM Trans. Graph. **28**, 43 (2009)
8. Bukhari, F., Dailey, M.N.: Automatic radial distortion estimation from a single image. J. Math. Imag. Vis. **45**(1), 31–45 (2013)
9. AlemánFlores, M., Alvarez, L., Gomez, L., SantanaCedrés, D.: Automatic lens distortion correction using one-parameter division models. Image Process. Line **4**, 327–343 (2014)
10. SantanaCedrés, D., et al.: An iterative optimization algorithm for lens distortion correction using two-parameter models. Image Process. Line **6**, 326–365 (2016)
11. Rong, J., Huang, S., Shang, Z., Ying, X.: Radial lens distortion correction using convolutional neural networks trained with synthesized images. In: Lai, S.-H., Lepetit, V., Nishino, K., Sato, Y. (eds.) Computer Vision – ACCV 2016. LNCS, vol. 10113, pp. 35–49. Springer, Cham (2017). https://doi.org/10.1007/978-3-319-54187-7_3
12. Yin, X., Wang, X., Yu, J., Zhang, M., Fua, P., Tao, D.: FishEyeRecNet: a multi-context collaborative deep network for fisheye image rectification. In: Ferrari, V., Hebert, M., Sminchisescu, Cristian, Weiss, Yair (eds.) Computer Vision – ECCV 2018. LNCS, vol. 11214, pp. 475–490. Springer, Cham (2018). https://doi.org/10.1007/978-3-030-01249-6_29
13. Liao, K., Lin, C., Zhao, Y., Gabbouj, M.: DR-GAN: automatic radial distortion rectification using conditional GAN in real-time. IEEE Trans. Circuits Syst. Video Technol. **30**(3), 725–733 (2020). https://doi.org/10.1109/TCSVT.2019.2897984
14. Liu, Z., Yeh, R.A., Tang, X., Liu, Y., Agarwala, A.: Video frame synthesis using deep voxel flow. In: Proceedings of IEEE International Conference on Computer and Vision, pp. 4473–4481, June 2017
15. Bao, W, Lai, W.-S., Ma, C., Zhang, X., Gao, Z., Yang, M.-H.: Depth-aware video frame interpolation. In: Proceedings of IEEE Conference on Computer Vision and Pattern Recognition, pp. 3703–3712, June 2019
16. Jiang, H., Sun, D., Jampani, V., Yang, M.-H., Learned-Miller, E.G., Kautz, J.: Super slomo: High quality estimation of multiple intermediate frames for video interpolation. In: Proceedings of IEEE Conference on Comput. Vision and Pattern Recognition, pp. 9000–9008, June 2018
17. Niklaus, S., Mai, L., Liu, F.: Video frame interpolation via adaptive separable convolution. In: Proceedings of IEEE International Conference on Computer Vision, pp. 261–270, October 2017
18. Liang, X., Lee, L., Dai, W., Xing, E.P.: Dual motion GAN for future-flow embedded video prediction. In: Proceedings of IEEE International Conference on Computer Vision, pp. 1762–1770, October 2017
19. Liu, W., Luo, W., Lian, D., Gao, S.: Future frame prediction for anomaly detection—a new baseline. In: Proceedings of IEEE Conference on Computer Vision and Pattern Recognition, pp. 6536–6545, June 2018
20. Li, Y., Fang, C., Yang, J., Wang, Z., Lu, X., Yang, M.-H.: Flow-grounded spatial-temporal video prediction from still images. In: Ferrari, V., Hebert, M., Sminchisescu, C., Weiss, Y. (eds.) Computer Vision – ECCV 2018. LNCS, vol. 11213, pp. 609–625. Springer, Cham (2018). https://doi.org/10.1007/978-3-030-01240-3_37
21. Khan, Z.A., Beghdadi, A., Kaaniche, M., Cheikh, F.A.: Residual networks based distortion classification and ranking for laparoscopic image quality assessment. IEEE Int. Conf. Image Process. (ICIP) **2020**, 176–180 (2020). https://doi.org/10.1109/ICIP40778.2020.9191111

22. Khan, Z.A., et al.: Towards a video quality assessment based framework for enhancement of laparoscopic videos. In: Medical Imaging 2020: Image Perception Observer Performance and Technology Assessment, vol. 11316, pp. 113160P (2020)

23. Aldahoul, N., Karim, H.A., Tan, M.J.T., Fermin, J.L.: Transfer learning and decision fusion for real time distortion classification in laparoscopic videos. IEEE Access **9**, 115006–115018 (2021). https://doi.org/10.1109/ACCESS.2021.3105454

24. AlDahoul, N., Karim, H.A., Wazir, A.B., Tan, M.J.T., Fauzi, M.A.: Spatio-temporal deep learning model for distortion classification in laparoscopic video. F1000Research, **10**, 1010 (2021). https://doi.org/10.12688/f1000research.72980.1

25. Beghdadi, A., Bezzine, I., Qureshi, M.A.: A Perceptual Quality-driven Video Surveillance System. In: 2020 IEEE 23rd International Multitopic Conference (INMIC), pp. 1–6 (2020) https://doi.org/10.1109/INMIC50486.2020.9318122

26. Beghdadi, A., Asim, M., Almaadeed, N., Qureshi, M.A.: Towards the design of smart video-surveillance system. In: 2018 NASA/ESA Conference on Adaptive Hardware and Systems (AHS), pp. 162–167, August 2018

27. Leszczuk, M., Romaniak, P., Janowski, L.: Quality Assessment in Video Surveillance. In: Recent Developments in Video Surveillance. London, United Kingdom: IntechOpen (2012). https://www.intechopen.com/chapters/34502, https://doi.org/10.5772/30368

28. Muller-Schneiders, T., Jager, H., Loos, S., Niem, W.: Performance evaluation of a real time video surveillance system. In: IEEE International Workshop on Visual Surveillance and Performance Evaluation of Tracking and Surveillance 2005, pp. 137–143 (2005).https://doi.org/10.1109/VSPETS.2005.1570908

29. Meyer, M., Hötter, M., Ohmacht, T.: A new system for video-based detection of moving objects and its integration into digital networks. In: Proceedings of 30th International Carnahan Conference on Security Technology, pp. 105–110 (1996)

30. Beghdadi, A., et al.: Challenge session: ICIP 2022. VSQuAD. https://vsquad2022.aliqureshi.info/index.html. Accessed 05 Feb 2022

31. Dosovitskiy, A., et al.: An image is worth 16x16 words: Transformers for image recognition at scale. In: ICLR (2021)

Author Index

Printed in the United States
by Baker & Taylor Publisher Services

Printed in the United States
by Baker & Taylor Publisher Services